An Alien in Antarctica

An Alien in Antarctica

Reflections upon
Forty Years of Exploration and Research
on the Frozen Continent

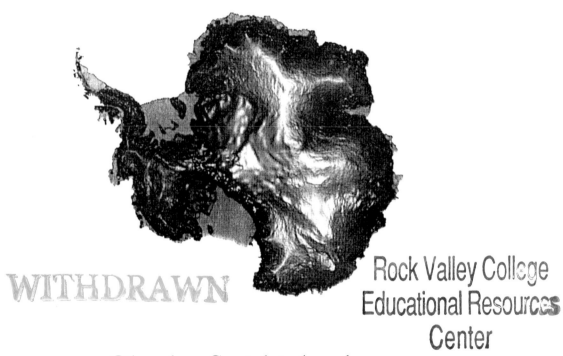

Charles Swithinbank

The McDonald & Woodward Publishing Company
Blacksburg, Virginia
1997

The McDonald & Woodward Publishing Company
P. O. Box 10308, Blacksburg, Virginia 24062-0308

An Alien in Antarctica

© 1996 by Charles Swithinbank

All rights reserved. First printing April 1997.
Composition by Rowan Mountain, Inc., Blacksburg, Virginia.
Printed in Canada by DWFriesens, Altona, Manitoba.

02 01 00 99 98 97   10 9 8 7 6 5 4 3 2 1

Library of Congress Cataloging-in-Publication Data

Swithinbank, Charles.
   An alien in Antarctica : reflections upon forty years of
exploration and research on the frozen continent / Charles Swithinbank.
     p.   cm.
   Includes bibliographical references (p.   –   ) and index.
   ISBN 0-939923-43-2 (hardbound : alk. paper)
   1. Swithinbank, Charles.   2. Explorers—Great Britain—Biography. 3. Antarctica—
Discovery and exploration.   I. Title.
G875.S95A3   1996
919.8'904—dc20   96-26746
                                                                                             CIP

Reproduction or translation of any part of this work, except for short excerpts used in reviews, without the written permission of the copyright holder is unlawful. Requests for permission to reproduce parts of this work, or for additional information, should be addressed to the publisher.

Contents

Preface		vi
Acknowledgments		ix
Prologue		xi
Chapter 1	Return to the Ice	1
Chapter 2	The Ross Ice Shelf	11
Chapter 3	Echoes of the Heroic Age	27
Chapter 4	Into the Unknown	41
Chapter 5	Mechanical Dogs	57
Chapter 6	Fatal Accident	71
Chapter 7	The Far South	83
Chapter 8	Armchair Glaciology	101
Chapter 9	Helicopters Unlimited	119
Chapter 10	The South Pole	131
Chapter 11	Runways on Ice	147
Chapter 12	Envoys from Outer Space	163
Chapter 13	The Acid Test	175
Epilogue		185
Acronyms		193
Glossary		195
References		199
Index		207

Preface

This is an account of one small part of a magnificent enterprise — the United States Antarctic Program. American activities in the Antarctic have spanned 40 years and are now more valuable than ever. Scientific results exceed — by a large margin — those of any other nation. While accounts have been written about life at coastal research stations, none has sought to describe the modern face of geographical exploration. This book is not about science, but about adventures in the pursuit of science. I had the privilege of taking part in six inland expeditions in Antarctica between 1959 and 1989 with a variety of objectives. The only common thread is that all were concerned with the study of ice.

My career within the US program spanned a period of unprecedented change. Schooled, as I was, in the British tradition of traveling the hard way, with man-hauled sledges, dog teams, and spartan rations, my move to the US plunged me into a world of giant icebreakers, aircraft, tractors, gourmet food, and a US Naval Task Force struggling to come to terms with the eccentric ways of scientists. Over 30 years I crossed paths with many an unsung hero and to them I dedicate this story.

The title of the book derives from the US Immigration and Naturalization Service, which describes any foreigner working in the United States as an *alien*. In Antarctica, where hundreds of miles of virgin snow separate neighbors, humans sometimes feel like aliens.

Charles Swithinbank
Cambridge, England

Acknowledgments

First, I have to thank Jim Zumberge and the University of Michigan, Ann Arbor; the National Science Foundation, Washington, DC; the US Naval Support Force, Antarctica; the US Geological Survey, Reston, Virginia; the US Army Cold Regions Research and Engineering Laboratory, Hanover, New Hampshire; the Scott Polar Research Institute, University of Cambridge; and the British Antarctic Survey — also in Cambridge — for 30 years of support and forbearance, without which none of the events described in this book would have come to pass.

The following people were kind enough to read and comment on drafts of parts of the work: Colin Bull, Dick Cameron, Bob Dale, Bob Headland, Terry Hughes, Michele Raney, Gordon Robin, Valerie Sloan, David Socha, Mary Swithinbank, Kay Tate, Tom Taylor, Edith Taylor, Dan Weinstein, and Bob Wells. Guy Guthridge and Nadene Kennedy of the National Science Foundation helped with the rendering of names.

For photographs and permission to use them I thank the US Geological Survey (pp. 19, 29, 33, 62, 66, 86, 107, 112), US Navy (pp. 6, 17, 25, 28, 50, 78), US Navy courtesy of David Burke (p. 47) National Science Foundation (p. 48), and Terry Hughes (p. 134). Jonathan Bamber supplied the ERS-1 satellite composite on the title page and Bärbel K. Lucchitta of the US Geological Survey supplied the digitally-enhanced NASA Landsat image on p. 37.

Finally, no family could have been more supportive than my wife Mary together with Anne, Carol, and Kelvin. They encouraged me and kept a happy home throughout my long absences in Antarctica. To them I owe a boundless debt of gratitude.

Prologue

I still have bad dreams about it. From the copilot's seat in the lead helicopter I could see that the eight-mile-wide glacier flowed through a canyon between precipitous cliffs 1,000 feet high. At the controls beside me was Commander Manson Krebs of the US Naval Support Force, Antarctica. A chubby, genial man, his hair was greying from long years as a pilot in war and peace. Flying at a height of 50 feet above the ground, he always seemed more at ease close to mother earth. "To conserve fuel," he said. The second helicopter — or "helo" as the Navy called them — with Lieutenant John Hickey, USN, in the driving seat, flew close behind us with my colleagues Tom Taylor and Dave Darby.

The unnamed and unexplored glacier that we now looked down on was riddled with crevasses. The very idea of landing on it took my breath away. After unloading the camping gear on the rim of the canyon, Darby and I took off with Krebs to set up a line of markers across the glacier as part of a plan to find out how fast it moved. Our markers — stakes we called them — were 10-foot lengths of aluminum irrigation pipe painted with bright red stripes and topped by a large red flag. Krebs was momentarily blinded by snow flurries as we took off; the helo bounced hard on the lip of the valley and my heart skipped a beat as we plunged over the edge. Like many another backseat driver, I tried to fight off a feeling of helplessness.

Our plan was to plant the stakes at one-mile intervals. But the ice was scarred by such a confusion of snow-bridged crevasses that not even the helo could find space to land safely. The first spot was in a really dangerous area but Krebs asked if it was safe to land. Emphatically, I said "No!" The surface was rough, and it was impossible to tell where it was solid and where treacherous.

The alternative was to use the helicopter's rescue winch, designed, perhaps, for rescuing downed airmen from the sea. With this device I could be lowered to the ice while the helicopter hovered above. All I had to do was wriggle into a form of harness and slide off the edge of the door. I was not unprepared for this situation; nevertheless, the events of the next few minutes served to dampen my enthusiasm for some aspects of glacier study.

The winch wire looked awfully thin. Dangling between earth and sky, revolving like a spider suspended from a strand of its web and buffeted by the rotor blast, I recalled the words of Captain Scott when — in February 1902 — he made the first balloon ascent in the Antarctic, and

"... felt some doubt as to whether I had been wise in my choice."¹ Once my feet were on the snow, I ducked out of the harness and signalled Krebs to come down and hover just off the surface. Then I took hold of the ice drill and stake that Darby handed down from the cabin. Meanwhile Krebs kept his eye on a rear-view mirror mounted outside the cockpit, watching my fumblings below and behind him.

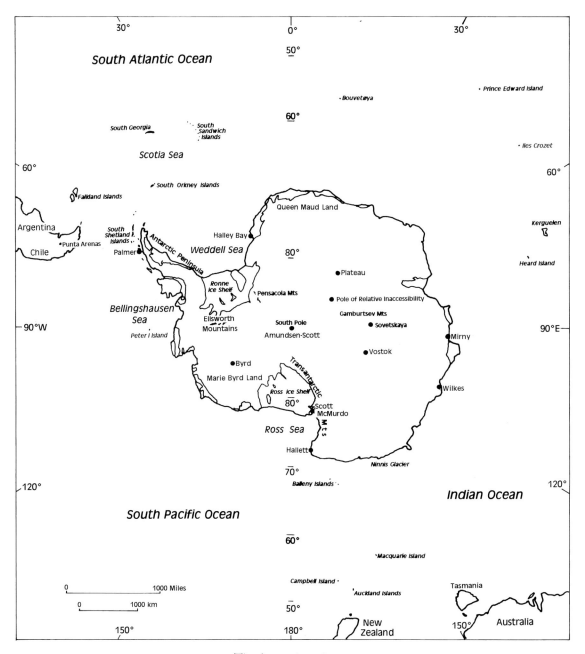

The Antarctic regions.

I waved him away, quickly made a hole with the ice drill, and planted the pipe in it. My troubles only really began when Krebs returned to maneuver the helo sideways towards me. He slowed, for understandable reasons, with my newly established survey marker safely beyond the sweep of his rotor blades, and beckoned to me with his head. I considered it tantamount to suicide to walk on that surface, so I beckoned to him with my head. Nothing happened. I did not want to be left on the glacier, so I took a step forward on to what appeared to be hard snow. Both my feet disappeared through a snow bridge. But the ice drill straddled the crevasse and I quickly climbed out. At this point Darby threw me the end of an alpine rope. I had to take off both gloves to tie it round my waist. Fortunately there were few witnesses to the spectacle of a shaken scientist advancing on hands and knees at the end of a rope belayed to a doorpost of the helicopter. With relief I clambered inside to relative safety. Sharp ice crystals had lacerated my hands. First-aid consisted of putting on gloves to hide the blood.

The surface was not so bad in the middle of the glacier and Darby jumped out to help with the drilling. Krebs rested the wheels of the helicopter lightly on the snow while taking most of the weight on the rotors. Each time we had to remember, on penalty simply of our lives, not to allow one end of the long pipe to tilt up to the rotor blades while it was being unloaded. We had visions of the salami-like slices to which we might be reduced if Krebs placed his wheels on a weak snow bridge. Half an hour passed, and we clambered back into the cabin after setting up the stakes that we had planned.

The next glacier also had no name, though later it was named for the legendary American explorer Richard Evelyn Byrd (1888–1957). As soon as I saw it I realized that here was a glacier possibly unparalleled on our planet. Ice was streaming towards the ice shelf from an outlet 20 miles across — as wide as Long Island Sound. The glacier surface was like an ocean in torment. As far as the eye could see, serrate ridges and giant furrows were aligned with the abrupt rock walls of the valley. We could follow stream lines sweeping uninterrupted through 80 miles from the high plateau inland down to the ice shelf. A backdrop of rugged mountain ranges added to the stark, prismatic beauty of the scene. The scale of the whole landscape was overwhelming.

The task of hovering over the ice this time fell to John Hickey. We dropped into the valley and flew low with mounting apprehension. When the moment came for the high wire act, as we called it, Hickey lowered me gently to the ice at the end of 20 feet of wire. He could not hover too low for fear of touching the rotor blades on an ice pinnacle, but he knew how to hold the machine steady. In two places he unwittingly lowered me on to a snow bridge, so that instead of ducking free of the harness I found myself gesticulating wildly at his calm, clean-shaven image in the rear-view mirror. Most of the surface consisted of steep and slippery bare ice, making it necessary to work without moving my feet to avoid careening into a hollow. After I had drilled in each stake, Hickey had to maneuver the swinging harness into my arms. This he managed with consummate skill in spite of a strong wind blowing down the glacier. After perhaps an hour, there were seven stakes in a line across the glacier.

That was 22 November 1960. There have been other moments of stark terror in the course of my 46-year involvement in Arctic and Antarctic research and I resolved — someday — to write about them. But where to begin? Could I explain how I arrived at this point? What was my purpose? What led a fairly normal young man to find such an abnormal occupation? It is a long story — so I shall tell it only briefly.

My mother was to blame. When I was eight years old she read aloud stories of adventure in far-off lands, of explorers hacking their way through jungle or dragging their sleds across virgin snow. The message she sought to instill — and she repeated it often enough — was: "Don't get stuck in an office like your father." My father was a classical scholar, educated at Eton and Balliol College, Oxford, who spent his professional life as a District Commissioner in Burma. Burma, now Myanmar, was at that time part of what the British referred to as their Indian Empire.

My mother's judgment was unjust, because my father did more traveling through his realm than any of his contemporaries. But he still spent most of his time in an office.

In Burma of the 1920s, my mother was the only woman who could wield an elephant gun. Not for sport but for protection. She was a naturalist who loved animals. She told of swinging her tame leopard by its tail into a tree. Fortunately the leopard, too, thought this was fun. At the time I was unaware that my mother's love of travel and open spaces must have rubbed off on her children.

In 1933, at the age of seven, she took me and my sister Jane to England for schooling. From then on I only saw my father at rare intervals when he was on leave from Burma. Primary schooling took six years and was that of a normal middle-class English child of the period.

As far as Britain was concerned, World War II began on 3 September 1939 when German troops invaded Poland. In the same week I went as a boarder to a public school, which is the British euphemism for a private school. This was Bryanston School, in Dorsetshire, where I spent five happy years. My academic record was undistinguished but I spent a lot of time in the fresh air dreaming — and reading — about adventures in faraway places. On three occasions I came close to being expelled from the school — for smoking and drinking, which were against the rules, and for making bombs, which was against the law.

The family home was in Kent, 30 miles southeast of London. During school holidays I watched squadrons of German bombers passing overhead on their way to blast London. Sometimes they were beaten back by British fighters and jettisoned their bombs around us. I delighted in jumping into fresh craters to recover still-warm fragments of the bombs.

At night we sheltered in the family's empty fishpond, which in peacetime doubled as a swimming pool. My mother built a roof over it, added a foot of topsoil and covered the whole with turf to make it look like an innocuous grassy mound. During air raids, shrapnel from anti-aircraft shells often rained from the sky. To protect ourselves on the nightly pilgrimage to the pool, we each held on our head one of the thicker volumes from my father's library. We hoped it would be impenetrable.

In June 1940, Jane was evacuated to Canada for safety. But my own baptism of fire brought only a longing to get into the fray and beat hell out of the Germans in any way possible. At school, I joined the Air Training Corps, and after being allowed to take the controls of an aircraft for a few minutes, decided that the Royal Air Force was for me. They told me to wait until I was 18½ years old. By 1944, however, it was becoming obvious that the war might be over before I reached that age. So at age 17½ I left school and joined the Royal Navy. That was in the same week as D-Day (6 June 1944), when the allied forces invaded France.

After training as an Ordinary Seaman, I was promoted to Midshipman and served for the next 2½ years in a cruiser, an aircraft carrier, and a minesweeper. A visit to Spitsbergen in 1945 left a lasting impression: the spectacular Arctic landscape, the clean air, and the pioneering spirit of the settlers made me feel I belonged there. I vowed to come back. The most valuable spin-off of the Navy years was a training in navigation that was to stand me in good stead when, years later, I had to navigate everything from dog-sleds to aircraft.

Leaving the Navy in October 1946 as a Sub-Lieutenant, I gained entrance to the University of Oxford to read for a degree in geography. I had no idea where this would lead. During my final year I still had no idea, and to fill in time proposed to join an expedition aiming to map the glaciers of Mount Kenya. Then at an Exploration Club coffee party one day, Scott Russell, a botanist on the teaching staff and an accomplished mountaineer, casually asked if I would like to go to the Antarctic. It took me a fraction of a second to reply. In that second the course of my life was changed for ever.

Scott Russell had been asked to look out for possible recruits for the planned Norwegian-British-Swedish Antarctic Expedition of 1949–1952. They needed an assistant to study ice. Having spent one summer with an undergraduate expedition trying to cross the Vatnajökull Ice Cap in Iceland, and another with an expedition to the Gambia River in west Africa, I was one of very few applicants with an expedition record. From the Iceland expedition I already had research papers accepted for publication in scientific journals. But I was only 22 years old and very much a beginner in glaciology — the study of ice. This hardly mattered because I would be working under an experienced Swedish glaciologist, Dr. Valter Schytt of the University of Stockholm.

Psychologists sometimes ask me whether going on an expedition was an escape or a pursuit. I never thought about it. In later years I did meet people on expeditions who were escaping from women, debt, or the law. But for me it was a vocation — or perhaps an unconscious attempt to live up to my mother's admonition to keep out of an office.

In 1949, ours was an international expedition supported by the governments of Norway, Britain, and Sweden. The group consisted of six Norwegians, four Swedes, three Britons, a Canadian, and an Australian — all of us male. One of the

Swedes and three of the Norwegians were married, but I was a bachelor and remained so until 1960. The plan was to spend two years in Queen Maud Land in the South Atlantic sector of Antarctica. The terrain was unknown and unmapped, though parts of it had been photographed from the air by a German expedition in 1939. Our principal scientific objectives were in the fields of meteorology, geology, glaciology, and mapping.

The full story has been well told elsewhere.[2] The expedition book was published in 10 languages; it was so good that in 40 years none of us has tried to better it. On 26 October 1949, I was one of five members of the expedition to sail from Sandefjord, Norway, on board the whaling factory ship *Thorshøvdi*. The expedition's own ship *Norsel* was much smaller and had nowhere to house sledge dogs. *Thorshøvdi* had ample space for our 50 dogs. Besides looking after the dogs, our job was to prepare the expedition's three "Weasel" tractors for their ordeal by ice.

The remainder of the expedition sailed a month later in *Norsel*, a 600-ton Norwegian sealer. For a month we — the whaling contingent — watched as up to a thousand ton of whales were sliced up daily and consigned to boilers below. After 11 weeks we met up with *Norsel* and moved from the big whaler to the small sealer in the lee of an iceberg.

After much searching along the icy coast of Antarctica, we established our base on a floating ice shelf that we devoutly hoped was firmly attached to the mainland. Icebergs periodically calve from ice shelves and there was no guarantee that we might not one day find ourselves drifting out to sea. There was no better site, so it was a risk that had to be taken. We unloaded onto the ice everything needed for a two-year stay. This included a third year of food and fuel because there was no certainty that a ship could get through the pack ice every year.

Southwards lay a monotonously level plain reaching to a line 20 miles away where the ice surface rose steeply towards a horizon perhaps 1,500 feet above sea level. Nowhere, not even to the farthest horizon, was there any sign of rock showing above the ice; just whiteness in directions east, west, and south — and an ocean of pack ice to the north.

Norsel left us after 10 days and steamed north. From then on, our little band of 15 lived, if not on *terra firma,* at least on a thick layer of ice, for two long years. We spent many months exploring inland, mapping, studying the ice and the geology of mountains that we found up to 350 miles inland. Three lives were lost in a drowning accident, and Alan Reece, one of two geologists, lost an eye after a chip of rock flew into it. The rest of us were lucky. We had narrow escapes but lived to tell about them. Perhaps because of that, and the challenge of overcoming difficulties, this was a most rewarding period of my life.

Two years later, in the first week of January 1952, *Norsel* reappeared to carry us home. This was my last view of Antarctica for eight years. I spent four of those years writing up the scientific results of the expedition for a doctorate at the University of Oxford. Everything was published as one volume in the expedition's scientific results series.[3] I spent the next four years in Cambridge studying the distribution of pack ice in the Northwest Passage on behalf of the Defence Research Board of Canada. This was mainly a literature search, but in the summer of 1956 it did give the opportunity to circumnavigate Baffin Island in Canada's newest icebreaker HMCS *Labrador*.

The city of Cambridge has been the hub of British polar exploration and research for 75 years. The University of Cambridge, founded in the 13th century, oversaw the establishment of the world's first polar institute in 1920. The Scott Polar Research Institute was named for the British explorer Captain Robert Falcon Scott (1868–1912) who died on the return journey from the South Pole. It houses the world's largest polar library and has staff actively involved in Arctic and Antarctic research. Later, Cambridge became also the headquarters of the British Antarctic Survey (BAS). BAS is a government organization with a staff of more than 400; it operates two ships, five aircraft, and four permanent research

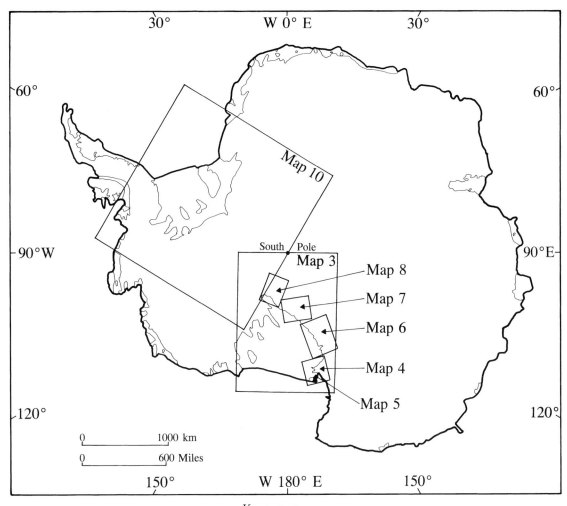

Key to text maps.

stations in the Antarctic.

While doing the Arctic work I was invited to return to Antarctica, first by a British, then by an American, and finally by an Australian expedition. But I had signed on for the Arctic, so that was that.

By 1959, I had been earning a living from polar research for 10 years. I was qualified for nothing else. But few people had ever made a career out of polar work. Could I survive without changing direction?

I made up my mind to find out.

---

[1] R. F. Scott. *The Voyage of the Discovery* (2 vols). London, Smith Elder & Co., 1905 (Vol. 1, p. 199).

[2] John Giaever. *The White Desert. The Official Account of the Norwegian-British-Swedish Antarctic Expedition.* New York, E. P. Dutton & Co., 1955.

[3] Charles Swithinbank. *Norwegian-British-Swedish Antarctic Expedition 1949–52, Scientific Results*, Vol. 3. Oslo, Norsk Polarinstitutt, 1957–1960.

Chapter One

Return to the Ice

Beyond this flood a frozen Continent
Lies dark and wilde, beat with perpetual storms
Of Whirlwind and dire Hail, which on firm land
Thaws not, but gathers heap, and ruin seems
Of ancient pile; all else deep snow and ice.
John Milton (1608–1674)

Pubs in Britain are good places to look for beer and *bonhomie*. It was in February 1959 that I was on a pub-crawl in Cambridge with Jim Zumberge. Pub-crawling involves drinking at three or more pubs in a row. We started at The Pickerel, then ambled over Magdalene Bridge to The Baron of Beef. The walls were hung with watercolors of American bombers, showing that the Baron was a popular hostelry with US servicemen during the war. Later in the evening, meandering towards The Kings Arms, Jim realized that we had a common interest in Antarctica and a yearning to return there. By the time we staggered into The Red Cow, we felt like old friends.

Little did I realize that this was the start of a 30-year association with the US Antarctic Research Program, and that it would lead to a series of unprecedented journeys through some of the most magnificent and pristine wilderness areas of our planet. I have set out my personal story in the following pages.

Zumberge was a Professor of Geology at the University of Michigan in Ann Arbor. He had been to the Antarctic in 1957–1958 to study the Ross Ice Shelf and proposed to go again. A bright, personable, and athletic man who looked set to go a long way in life,[1] he said over our fourth pint of beer, "Charles, if you ever need a job, let me know."

"Thank you," I replied, and thought no more of it.

Seven months after the pub-crawl, when the work I was doing for the Canadian government was finished and on the way to publication, I had reason to wonder whether Professor Zumberge might confirm or deny my hazy recollection of his invitation. The world would be a sorry place if people were held accountable for off-the-cuff remarks in a pub. But he did remember. I phoned, and the response was immediate — Jim was never one to dither. He said "Yes, if I can raise some money."

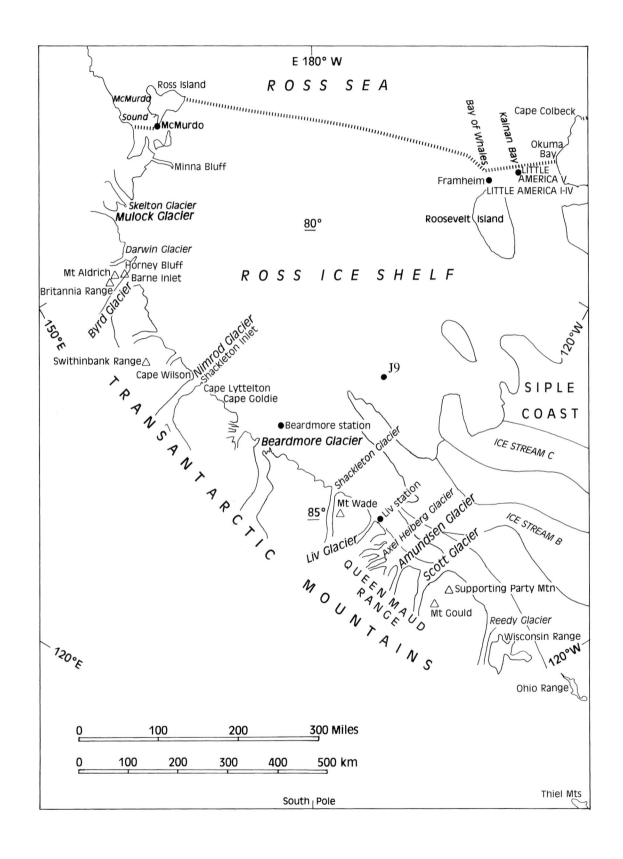

He advised me to contact the right person in Washington.

That right person was Albert P. Crary, Chief Scientist for Antarctic Operations at the National Science Foundation (NSF). NSF controlled the funds for US Antarctic work, and Zumberge could not spend anything without Crary's approval. Crary himself was answerable to the Director of the Office of Antarctic Programs, Thomas O. Jones. Fortunately, I had met Crary — we knew him as Bert — at several Antarctic science conferences. He at once phoned Zumberge to let him know that money could be found to employ me. Jim cabled an invitation and I phoned back.

"When can you come?" he asked.

"Anytime you like," I said.

He added simply, "It will take me a few days to rustle up the money. Come over when you can."

I did, and we collaborated happily for years afterwards. Zumberge had been studying the deformation of the Ross Ice Shelf in the hope that it could serve as an analog to the deformation of rocks. He planned to return to make a series of measurements on another part of the ice shelf. We hoped to discover whether more, or less, snow was falling on the ice sheet than was being lost into the surrounding ocean by melting and by the calving of icebergs. In other words, was the ice sheet in balance with the prevailing climate, or instead growing, or perhaps shrinking? We were to study these things in the hope of understanding how ice sheets respond to climatic change.

Antarctica is the only contemporary example of a continent in the grip of an ice age. On other continents, ice ages have come and gone — and it is generally believed that they will come again. To those who asked what point there was in studying such esoteric matters, we would reply that the Antarctic ice sheet is the principal factor controlling sea level throughout the world. Sea levels have fluctuated by hundreds of feet in past millennia, and today, millions of people live in cities and countries barely above today's sea level. What is in store for them?

Zumberge's plan was to traverse westwards with a couple of Sno-cat tractors from Kainan Bay (latitude 78° 16' S, longitude 162° 28' W) to measure the rate of movement of the ice shelf. Our final destination was the main US base built in 1956 on Ross Island beside Captain Scott's *Discovery* expedition hut in McMurdo Sound. Scott's hut was still standing 58 years after it was built. Between the Bay of Whales and McMurdo — as the present American station is known — lay 450 miles of flat, almost featureless snowfield representing the surface of a floating ice sheet approaching the size of Texas and in places up to 3,000 feet in thickness.

We left Ann Arbor on 27 November 1959. I did not conceal my excitement at once more heading for Antarctica — the least-explored place on Earth. At Christchurch in the south island of New Zealand we were met by Edward Goodale, the local liaison officer for the National Science Foundation. Eddie was a real veteran who had wintered at the very first

Ross Sea to the South Pole.

"Little America" station of Richard Evelyn Byrd's 1928–1930 Antarctic expedition.

The next day Jim introduced me to another veteran, the New Zealander Sir Edmund Hillary,[2] who together with Sherpa Tenzing in 1953 had made the first ascent of Mount Everest.[3] Later he was leader of the New Zealand Antarctic expedition in support of Sir Vivian Fuchs' Trans-Antarctic expedition of 1955–1958.[4] We listened with interest to the plans of this giant of a man for another visit to the Himalayas.

While we were in Christchurch the radio carried the news of an historic development in the political status of Antarctica. A treaty was signed in Washington, DC guaranteeing free access to the continent for any peaceful purpose. The signatories of The Antarctic Treaty, as it is known, were the governments of Argentina, Australia, Belgium, Chile, France, Japan, New Zealand, Norway, South Africa, the Soviet Union, the United Kingdom, and the United States. Each of these countries had recent experience of collaborating in the cause of science during the International Geophysical Year of 1957–1958. Harmonious relationships were continuing in the Antarctic despite the stresses of the Cold War elsewhere.

Old icebergs in open pack ice.

The need for a treaty arose from a common interest in preventing the militarization of Antarctica, and the need to accommodate differing views on sovereignty. Seven of the signatory nations had made claims to sectors of Antarctica and three of the sectors overlapped. Reconciling these differences was a brilliant feat of diplomacy, and the treaty laid the groundwork for cooperation that has stood the test of time.

Christchurch has a delightfully English atmosphere about it. The style of buildings, the pace of life, the park by the river in the middle of town — all these things reminded me of Cambridge in England, or Stratford-upon-Avon. Christchurch's river is also the Avon. The nearest seaport is Lyttelton, five miles away within the rim of a flooded volcanic caldera. There we boarded the USS *Arneb* on 5 December. Polar clothing had been shipped from a warehouse in Washington, DC and the rest of our equipment was already on board.

Jim Zumberge, at age 36, was my senior by three years; we were both older than most of the scientists bound for Antarctica. Jim had brought our small party up to strength with Jim Schroeder, a surveyor from Port Hueneme, California, and Fred Jacobi, a rural schoolteacher now masquerading as tractor mechanic. Jacobi was from Cleveland, Wisconsin, and had worked in Greenland with the US Army. Like many others, he came south for a change of scene — almost a vacation. Other scientists on board were Tony Gow (glaciologist), Lin Gressitt (entomologist from Hawaii), and Hugo Neuburg (cosmic ray physicist). Major Antero Havola of the US Army was berthed with us because he was "in the wrong service," as one Naval officer rather tactlessly put it.

As in all armed forces, there was an order of precedence defined by rank. Being without rank, the civilians on board presented a problem. We were informed that those of us with a Ph.D. were to be equated with Lieutenant-Commanders, and those with a B.A. or B.Sc. degree were Lieutenants. It meant that we all belonged in the wardroom and had to dress accordingly. Places at table were assigned and napkin rings provided.

In contrast to most governments, the US has used military personnel and equipment to support its scientific programs. The scale of the American operation was so large that no country could have done it without resorting to the military. In the 1950s, there was a shortage of men with Antarctic experience. Many of Byrd's generation were no longer available. The advantage of using the Navy was that ships and aircraft could be temporarily deployed to the Antarctic but then returned to normal service during the southern winter (the Northern Hemisphere summer). The icebreakers generally worked in Arctic waters from July to September and in the Antarctic from December to March.

The drawback of using the military is that they deploy many more people than a civilian operator would use for the same job. The result can be an imbalance, with scientific personnel feeling overwhelmed by the size — and what seems to them the inertia — of the military system. The two cultures felt ill at ease with each other. But having served in the Royal Navy, I had some understanding of the military mind. It was counter-productive to remind them that science was the *raison d'être* for their presence in the Antarctic. Our strategy was to get them interested in what we were trying to do. Once we had generated some enthusiasm, there were many occasions on which they helped above and beyond the call of duty.

Arneb was named after the constellation of that name. An ice-strengthened "attack cargo ship" displacing 14,200 ton, she was 460 feet long and had a steam turbine engine of 6,000 horsepower that could drive her along at 16 knots. She had earned battle honors in the Pacific war in 1944 and 1945, and in the 1955–1956 Antarctic summer served as flagship on Operation Deep Freeze I. Deep Freeze was the Navy's name for a massive construction program to establish not only McMurdo Station but also Little America V as well as other stations. Besides the few passengers, *Arneb* was carrying 4,400 ton of cargo, most of it for McMurdo Station.

The name Little America has an illustrious history. The first Little America was the winter base of Byrd's first Antarctic expedition in 1929.[5] Little America II was the winter base of the second Byrd

Planning meeting on board USS *Arneb*. Back row: Tony Gow, Fred Jacobi, the author; front row: Jim Zumberge, Antero Havola, Hugo Neuburg, Lin Gressitt.

The face of a tabular iceberg — about 80 feet high and showing annual snow strata.

expedition in 1934.[6] Little America III was the winter base of the US Antarctic Service Expedition in 1940.[7] Little America IV was the summer camp in 1946–1947 of the US Navy Operation Highjump.[8] Byrd led all four expeditions. These Little Americas were all at the Bay of Whales, close to the site of the Norwegian explorer Roald Amundsen's base Framheim, which was occupied in 1911.[9]

A reconnaissance in 1954–1955 by the icebreaker USS *Atka* showed that the Bay of Whales had changed so much that it was unusable. Ice cliffs 80 feet high had taken the place of low ramps used by Amundsen and Byrd to unload their ships. Kainan Bay, 45 miles farther east, offered a good site, and here Little America V was established and occupied during the winters of 1956, 1957, and 1958. In 1959, besides McMurdo and Little America V, the US was operating two other stations: South Pole at 90° S and Byrd at 80° S, 120° W. In addition, there was Hallett, a joint US/New Zealand station (72° 18' S, 170° 18' E). Two former US stations had been taken over by foreign governments but still had a few Americans wintering with them: the Australian Wilkes Station (66° S, 111° E), and the Argentine Ellsworth Station (77° 43' S, 41° 08' W).

All the Little Americas were built on the Ross Ice Shelf. An ice shelf is a floating ice sheet that survives as a moving glacier in part because of the continuous accumulation of snow on its surface. Layers two or three feet thick accumulate each year, progressively burying bases and stores. Thus after a few years every one of the Little Americas was buried to the point where the buildings collapsed from the weight of overlying snow. This is in contrast to building on rock, where permanent stations can be established without fear of being buried under snow. Unfortunately, less than one percent of the area of Antarctica consists of bare rock, so the choice of sites is limited.

USS *Glacier*, the largest icebreaker in service anywhere outside the Soviet Union, joined us on 10 December at the edge of the pack ice. That day there were riotous ceremonies on deck to celebrate crossing the Antarctic Circle. King Neptune, in full regalia, looked on as all newcomers to the Antarctic were shaven and doused with ice water.

A cool breeze coming off the ice floes felt good on our faces. One almost believed the old-time whalers' claim that they could smell pack ice. A short time later we were in amongst ice floes, nudging them gently to clear a path. On the first day in close pack ice, several of us were looking over the bow to where, 300 yards ahead, *Glacier* was breaking a path for us to follow. *Arneb* was progressing nicely when *Glacier* was brought to a standstill by a pressure ridge. By the time our CO, Captain Edwin A. Shuman, USN, had the engines going full astern, our momentum kept the ship moving inexorably towards the icebreaker. There was a sickening crash of tearing steel as *Arneb's* bow ploughed into the helicopter flight deck over *Glacier's* stern, and there were red faces on the bridge.

Luckily, damage to both ships was above the waterline. This sort of accident has happened before with ships in convoy through ice. It can be avoided by following the icebreaker at a greater distance, but then ice floes tend to drift into the broken channel behind the leader before the following ship can get past them. Like many before him, Captain Shuman learned the optimum distance the hard way.

There was an added embarrassment. The Navy task force commander, Captain Edwin MacDonald, USN, was using *Arneb* as his flagship at the time, so witnessed the collision. MacDonald was a colorful character with considerable experience of driving icebreakers. He never minced words in dealing with subordinates. Socially, he could be tactless but was good company on a one-to-one basis. In typically direct manner, he later published an article with the title "Our icebreakers are not good enough."[10]

We were sailing past a series of gigantic icebergs, their northward faces glinting in the sun. Icebergs more than 100 miles long have been seen in Antarctic waters. Most of those we saw were flat-topped, tabular bergs, but some were dome-shaped and crevassed all over, the so-called "breadcrust" bergs. They varied in size from a few hundred yards to 10 miles in length, a stately procession calved from some distant ice shelf, born to the sea and now borne by sea currents. Some, in all probability, would stay in Antarctic waters for a decade. Others would head north for warmer waters and so to their own demise.

Most Antarctic bergs like those we were passing contain so much air that they float high in the water and appear white — unlike Greenland bergs, which are closer to pure, bubble-free ice and there-

A cold day at the ice front.

fore denser and bluer. A 100-foot-high Antarctic iceberg may be supported by 400 feet of ice below sea level. It was quite staggering to calculate that even a small iceberg — by Antarctic standards — with dimensions of about one mile, can weigh a billion ton. No wonder there has been talk of towing a few of them to Saudi Arabia to supplement the water supply.

After a couple of days with the icebergs, while at the same time forcing our way through giant ice floes, we were joined by another icebreaker, USS *Atka*. As the British explorer Captain Sir James Clark Ross (1800–1862) found when he led his ships *Erebus* and *Terror* into these waters in 1841,[11] the pack ice is concentrated in a wide belt, and beyond it, there is commonly ice-free open water all the way to the coast. So it was with us. We came out of the ice at midnight on 12 December and reached Kainan Bay two days later. Fast ice — an unbroken sheet of sea-ice attached to the coast — filled the bay and we tied up alongside it to unload the cargo.

To celebrate our timely arrival, the crews of both ships held a beer-drinking party. The consumption of alcohol was forbidden on board ships of the US Navy, so the rule was circumvented by unloading the booze and having a party on the ice.

King penguins (*Aptenodytes patagonica*).

Sailors warmed their hands by burning the empty beer cases.

Having finished the tasks it was set up for, the camp at Little America V had been closed the previous summer. All the buildings, erected four years earlier on the snow surface, were now hidden under snow and we needed to do a lot of digging to find things. However, the radio masts still stood and there was a map of the camp that told us where to dig. Two Sno-cats and a great deal of food had been left the year before, and gradually we brought to the surface both vehicles (each one weighing 5,000 pounds), two ton of gasoline, and all the food needed for the journey. For the first time in our lives we could help ourselves to anything, knowing that if we left it, everything would be lost under the snow forever.

Four days after *Arneb* had put our party ashore, we were ready to start. In a brief but touching ceremony, we struck the last American flag to fly over the last Little America. I still have it.

---

[1] James Herbert Zumberge (1924–1992) was elected President of Grand Valley State College in 1962, and later, in succession, Dean of the College of Earth Sciences at the University of Arizona, Chancellor of the University of Nebraska-Lincoln, President of Southern Methodist University, and finally President of the University of Southern California.

[2] Edmund Hillary. *Nothing Venture, Nothing Win.* London, Hodder & Stoughton, 1975.

[3] John Hunt. *The Ascent of Everest.* London, Hodder & Stoughton, 1953.

[4] Sir Edmund Hillary. *No Latitude for Error.* London, Hodder and Stoughton, 1961.

[5] Richard Evelyn Byrd. *Little America. Aerial Exploration in the Antarctic, the Flight to the South Pole.* New York, G. P. Putnam's Sons, 1930.

[6] Richard Evelyn Byrd. Discovery. *The Story of the Second Byrd Antarctic Expedition.* New York, G. P. Putnam's Sons, 1935.

[7] Reports of the scientific results of the United States Antarctic Service Expedition 1939–41. *Proceedings of the American Philosophical Society,* Vol. 89, No. 1, 1945, pp. 1–398.

[8] Richard Evelyn Byrd. Our navy explores Antarctica. *National Geographic Magazine,* Vol. 92, No. 4, 1947, pp. 429–522.

[9] Roald Amundsen. *The South Pole. An Account of the Norwegian Antarctic Expedition in the "Fram", 1910–1912.* New York, Lee Keedick, 1913 (2 vols).

[10] Edwin A. MacDonald. Our icebreakers are not good enough. *United States Naval Institute Proceedings,* Vol. 92, No. 756, 1966, pp. 59–69.

[11] Sir James Clark Ross. *A Voyage of Discovery and Research in the Southern and Antarctic Regions during the Years 1839–43.* London, John Murray, 1847.

Chapter Two

The Ross Ice Shelf

> *This extraordinary barrier of ice, of probably more than a thousand feet in thickness, crushes the undulations of the waves, and disregards their violence: it is a mighty and wonderful object, far beyond any thing we could have thought or conceived.*
> — James Clark Ross, 30 January 1841

All the great explorers of this sector of Antarctica had traveled over the Ross Ice Shelf: Robert Falcon Scott (1901–1904 and 1910–1913),[1] Roald Amundsen (1910–1912),[2] Ernest Henry Shackleton (1907–1909),[3] and (1914–1917),[4] Laurence McKinley Gould (1928–1930),[5] and Richard Evelyn Byrd (1933–1935).[6] In 1957, US scientists led by Bert Crary had established that the ice is 500–3,000 feet thick and flowing under its own weight towards the Ross Sea. Gigantic icebergs calve and drift northwards from the 500-mile-long ice cliff that is its seaward margin.

The British explorer Sir James Clark Ross,[7] who discovered it in 1841, saw the seaward face — or ice front as we call it today — as a barrier to his ships. But throughout the 20th century the surface has served more like a highway — though a rarely used one. We felt a thrill of excitement on setting course for McMurdo, 450 miles away. Each Sno-cat was hauling two cargo sledges. Zumberge drove one vehicle and Jacobi the other, while Schroeder and I rode as passengers.

The first part of the journey was through a badly crevassed area. My companions had been reading NSF's *Survival in Antarctica*, which has a chapter on crevasses. "Crossing a crevassed area is always dangerous . . . there is no infallible method of detecting [them]," it said. "Snow bridged crevasses make travel very treacherous in . . . the margins and grounded areas of floating ice shelves."[8] We were in just such an area.

I have always been terrified of crevasses and did not feel at ease. Some of the snow bridges ahead were visible but others were not. At one point, quite unexpectedly, a violent lurch brought Jacobi's Sno-cat to a sudden halt. One end of his two-ton tractor had dropped through a snow bridge to reveal a menacing black hole beneath. The depth of some of these crevasses exceeded the height of a

Emperor penguins (*Aptenodytes forsteri*) come to greet visitors.

10-story building, so we felt lucky not to have dropped further. It took several hours of digging to recover the situation. Sledges had to be unlashed, unloaded one box at a time, and reloaded after moving the sleds. Throughout the operation, one false step could have spelt disaster.

I often thought how useful it would have been to have a light aircraft with us. Whenever we came upon crevasses, we could launch the plane and spy out the terrain from above. Surface travelers cross thousands of snow-bridged crevasses without knowing it. Most of the snow bridges remain intact, but when one does break under the weight of a man or a sledge, lives are at risk.

A few miles beyond the crevasses and by prior arrangement, a US Navy UC-1 "Otter" aircraft fitted with skis landed beside us. Out jumped a short, stocky young man with a ready smile. It was Jack Long, who came to serve as our second mechanic. The pilot was Commander Manson (Buddy) Krebs, USN, whom I afterwards got to know well. He was equally at home flying fixed wing aircraft or helicopters. It was said of him that one day in his Otter, bored with its slow progress across the vast ice shelf, he began to read a newspaper. Deep in a story of scandal, he failed to notice that the aircraft was descending. He was rudely awakened by a sharp jolt as the aircraft hit the snow and bounced back into the air. Elsewhere in the world, or without skis, such an event could be disastrous. But not here. On the other side of Antarctica there is a dome-shaped feature bearing the name Touchdown Hills.[9] Evidently Krebs is not the only pilot to have gotten away with it.

Zumberge had been troubled by a bladder infection and it was getting worse. By radio the Navy doctors advised him to get to McMurdo as soon as possible in order to be flown out to a good hospital. Rightly concerned that the Antarctic wilderness was no place to be with a health problem, Jim climbed aboard the Otter to be whisked away to McMurdo Station and from there to New Zealand.

He was mightily disappointed at the thought of missing the rest of the traverse. We were unhappy at being deprived of his expertise and lively company. However, four was a safe number to travel with, so we were not otherwise concerned. I took over the leadership.

We were following a trail that had been laid the year before by a tractor party under the command of Lieutenant-Colonel Merle (Skip) Dawson of the US Army. This was not like any trail through a northern forest. Snow had obliterated everything except for bamboo poles sticking up at intervals of 300 yards. To establish how much snow had fallen, we measured the height of each one in

Crevasse accident. The rear end of a Sno-cat has dropped through a hidden snow bridge. The tattered black flag on its bamboo pole shows that we were right on the trail. There are open crevasses in the background.

Fred Jacobi in the wanigan.

passing. To save stopping the Sno-cat, one of us wore skis and was towed behind on a rope. By hauling ourselves ahead on the rope, then letting it slide through our gloved hands, we completed each measurement before again gripping the tow rope.

Every 20 miles, if the sky was clear, we stopped to take sun angles with a theodolite to calculate our position, and at the same time erected various markers to be remeasured by later travelers. Once a day we made radio contact with McMurdo, using a Morse key — like any ship at sea — to report latitude and longitude. Indeed we were at sea, hundreds of miles from land, but separated from sea water by a thousand feet of ice. Although I had learned the Morse code while serving in the Royal Navy, I was now pathetically slow, but it hardly mattered.

We slept in a "wanigan" made from tarpaulins fastened over a rectangular wood frame mounted on one of the cargo sledges. It was crowded but cosy. Meals began by scooping up snow outside into a saucepan. Set over a Coleman stove, water-making was a slow process of adding further chunks of snow until the pot was full. We had brought along dried meat, pasta, biscuits, butter, assorted jams, potato powder, bacon, hamburgers, dried milk, canned fruit, cocoa, coffee, tea, and dried fruit juice. Stews were like no stews ever made before, the ingredients being a random selection based on the whims of the cook of the day. The fruit juice crystals were delicious and it took only a teaspoonful in a mug of water to believe that we were quaffing fresh orange or grapefruit juice.

The most memorable foods were "Bolton" rations. These were assorted prepacked portions made for campers. Heavy on carbohydrate but light on protein, fat, or anything to enjoy, we had a full belly after meals but felt hungry an hour later. Having found cases of Mars bars and Hershey's chocolate in danger of being buried forever at Little America, we had a surfeit of sweetmeats. But these, too, failed to stave off hunger for long. Generations before, explorers had learned that in cold climates, the body craves protein and fat as much as carbohydrate. Short-changing any one of them affects physical fitness. Luckily, ours was to be a short journey by Antarctic standards; we were lacking in stamina but felt no other symptoms.

How did we handle the call of nature? The short answer is "As rapidly as possible." We generally walked 50 yards downwind, dug a shallow hole in the snow and squatted over it. It helps to hold on to a shovel. There is no difficulty on a sunny day, but blizzards reward a slow performer with pants full of snow. There is a risk of frostbite in awkward places. The worst form of purgatory is when a dangling penis, desensitized by cold, freezes to the ground. An ice axe may be needed to free it. Returning to the warmer wanigan, melting snow leaves soggy underwear. We

At the International Date Line, 30 December 1959. The author (left), Fred Jacobi, Jim Schroeder, and Jack Long. The lower panels record a 1958 visit by Major Havola's traverse party.

dreamed of flushing toilets in warm rooms.

Shortly before Christmas, a twin-engined R4D (the Navy version of the Douglas DC3 — also known as an LC-47) unexpectedly flew over us. A door opened in the side of the aircraft and two mailbags were thrown out. They hit the snow 50 yards away. Gathering them up, we found mail, a fruit cake, and a bottle of whisky. The plane circled, dipped its wings, and flew on into the gloom of an overcast day. This was the first of many occasions on which deeds conveyed far more than words. We were having a very lonely Christmas. The aircrew knew it, and flew for hours to show that we were not forgotten.

Throughout our journey the weather — by Antarctic standards — was balmy. Air temperatures were generally +20° F to +32° F during the day though colder at night. The coldest recorded was +2° F. It is the wind, however, that bites to the bone. A wind of only four knots carries away several times more body heat than still air. I had a table of the so-called wind-chill factor, which shows the increase in cooling power of moving air. At 30° F, a 13-knot wind makes the body feel as cold as it would in still air at 10° F. But we were well-dressed, and the highest winds we ever recorded were about 20 knots.

On 31 December we crossed the international date line. Antarctica is the only place in the world where a tractor can drive across the date line. Elsewhere you sail across it or fly over it. Starting from the North Pole, the line runs south through Bering Strait, with diversions to avoid all the Pacific islands. Here, however, the line was exactly on the 180th meridian from Greenwich. Dawson's party had set up a bamboo pole to mark the place. On one side of the pole there was an empty fuel drum with a painted sign TODAY; on the other side one marked TOMORROW.

The sun was always above the horizon though on many days we could not see it. I knew it was day when the sun was high in the sky and to the north of us. At night it was low in the sky and south of us. My weather observations tapped out to McMurdo show that we were not always basking in sunshine:

> *NGD* [To McMurdo] *de* [from] *NLA02* [our codename] *QTH* [my position] *78° 52' S, 172° 40' E. Ceiling zero, no horizon, wind 4 knots, poor surface definition.*

One of the Sno-cats broke down, and despite the combined efforts of Jack and Fred, we had to radio McMurdo for spare parts. Owing to low cloud and whiteout it took four flights to find us. Whiteout is something people do not really believe in until they have experienced it in an accident. Caused by multiple reflection between a snow surface and overcast cloud, the effect is to diffuse the daylight and obliterate all shadows. Yet dark objects can still be seen over long distances. On a uniformly white surface, shadows provide the only clue to scale and to where the surface lies in relation to the observer. Men have been known to march unwittingly over an ice cliff in a whiteout. Aircraft hit the snow while the pilot thinks he is at a safe altitude. Snow bridges over crevasses, inconspicuous at the best of times, become undetectable.

An overnight stop on the Ross Ice Shelf.

Without any scale factor, I once mistook a discarded match-box, 20 feet from the tent, for a sledge party half a mile away.

When eventually the Otter landed, to our surprise out jumped a pilot in the uniform of the Royal Air Force. It was Squadron-Leader K. A. C. (Dickie) Wirdnam. His command of an aircraft boldly marked U S NAVY was explained by a pilot-exchange agreement common among NATO armed forces. The copilot was Lieutenant James (Jim) Weeks, USN, whom I got to know later as an R4D pilot. They brought the parts we needed and the mechanics got to work.

After cheery farewells, Wirdnam started up and gunned the engine for take-off. Nothing happened — the skis were frozen to the snow. Having had this problem years before, I stretched out a climbing rope and, with one of us on each end, slid it under the aircraft's skis; then with a sawing motion worked the rope slowly from front to back of the skis. Dickie, still at full throttle, was hauling back and forth on his control column. Finally the machine broke free and returned to its element. This method almost invariably works, but pilots live in terror that clumsy landlubbers might allow the rope to wrap itself round a whirling propeller, with consequences too awful to

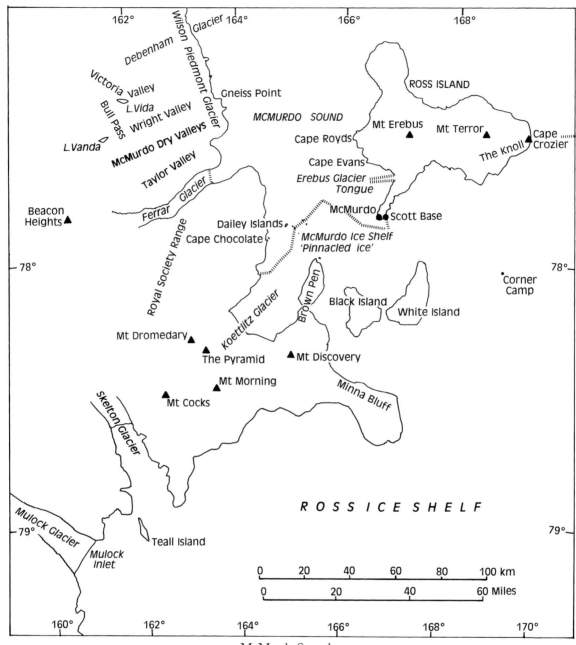

McMurdo Sound.

think about.

One day while taking sun angles with the theodolite, I noticed that the bubble in the spirit level of the instrument was slowly moving from one end to the other and back again. Though we could not feel any motion, it was evident that the 1,000-foot-thick ice we were camped on was being ever so gently tilted by a very long-wave ocean swell. Each complete cycle of the bubble took 280 seconds. People often wonder what imperceptible event triggers the calving of giant icebergs. Perhaps we had found the secret. But we had no wish to see the idea proven — at least not until we had moved on a safe distance.

On 10 January 1960 the two great volcanoes on Ross Island loomed over the horizon, the first bit of solid ground that we had seen since leaving New Zealand. Mounts Erebus and Terror were named by Sir James Clark Ross in 1841 for his two little ships. I felt a thrill of excitement because we were approaching the McMurdo Sound area where so many of the great dramas of the heroic age of exploration were played out. The British expeditions led by Captain Scott (1901–1904 and 1910–1913) and by Ernest Shackleton (1907–1909 and 1914–1916) all had their bases in McMurdo Sound. Their exploration dominated the history of this area from 1901 to 1917.

The next day we saw the Scott memorial cross on Observation Hill at a range of 25 miles. Crossing lines of crevasses near Scott's "Corner Camp," some black holes appeared as snow bridges collapsed

Ross Island from an altitude of 15,000 feet. The volcano Mount Erebus emits a plume of steam. Tractor tracks in the left foreground lead from a ship to the US McMurdo Station on Hut Point Peninsula (center). New Zealand's Scott Base is on the small promontory by the pressure ridges on the right-hand edge of the peninsula. Erebus Glacier Tongue (left middle distance) has serrated edges.

beneath the vehicles; fortunately nothing fell in. Crossing crevasses invariably brings on a feeling of foreboding — how long will our luck hold? But with stiff upper lips, we concealed these feelings from each other.

A few miles to the north, the runners of an upended 20-ton cargo sledge projected above the snow. We learned that the year before, a Caterpillar D8 tractor towing two 20-ton sledges, one loaded with explosives, broke through a snow bridge and fell into a crevasse. Both men in the cab of the tractor survived the fall and were rescued, but the tractor and sledges were abandoned because of the risk of another accident if attempts were made to recover them.

Triumphant at having coaxed two aged Sno-cats — abandoned at Little America because they were not worth saving — across 450 miles of ice shelf, we drove past the New Zealand government's Scott Base and over a low pass to McMurdo. We were greeted by George Toney, a former classical scholar who was serving as the NSF's representative at McMurdo. George, fair-haired and fiftyish, was a mild-mannered and very effective liaison officer who handled the sometimes delicate relationships between scientific staff and the Navy. His job was

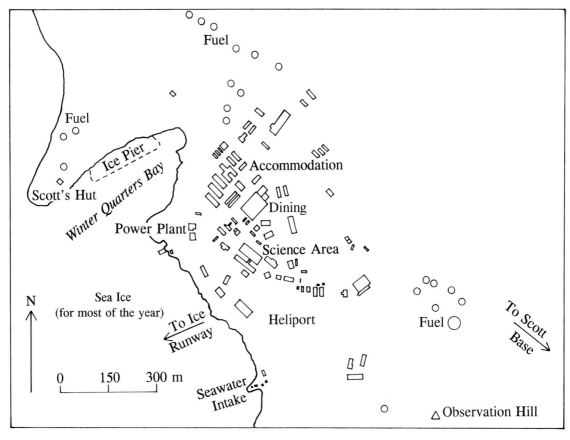

McMurdo Station.

not made easier by a few of the "USARPS," as we were called, who behaved like *prima donnas* in demanding more support from the military than it was reasonable to expect. George worked very long hours, knowing that part of his job was to socialize with the naval officers to help them understand the rationale behind our work. He slept heavily and needed three alarm clocks dangling from strings just above his head in order to wake in time for breakfast.

However, even George's charm failed us one day. The two cultures at McMurdo had firm ideas about property. USARP generally held the short end of the stick because the Navy controlled the whole logistic enterprise. USARP owned a few vehicles but was desperately short of equipment that could be considered its own. We felt, and George felt, that USARP was entitled to keep our two Sno-cats, which we had rescued from oblivion and driven all the way from Little America. The Navy pointed to the US Navy registration number on each one, and to our cost, insisted that right was on their side. They owned them, so they kept them.

By Antarctic standards, McMurdo was a city. Ever since 1956 it has been, and still is, the largest settlement on the continent. Ramshackle and desperately untidy, it resembled a frontier town that had grown up with no attempt at planning. Where Shackleton's Siberian ponies trod 50 years before, Navy trucks pounded along dirt roads that led nowhere except down to the sea-ice and a 10,000 ton cargo ship unloading onto a string of sledges. Giant bulldozers had scarred the landscape. One or two mean streets were detectable, perhaps because the first bulldozers drove that way and nobody in their right mind would erect buildings in the path of a bulldozer. There was a little church, "Our Lady of the Snows," built of corrugated steel with a tiny wooden steeple complete with bell to summon the faithful. Helicopters buzzed overhead. Four R4D aircraft were parked beside a runway on the sea-ice. Husky dogs wandered the streets begging for attention or a morsel of food.

Subject to no supra-national authority, Antarctica has no agreed rule of the road. At McMurdo, vehicles drove on the right. A mile away, at Scott Base, vehicles kept to the left because that is the rule of the road in New Zealand. One summer, a road was constructed to connect the two stations. It is related that the rule for this part of Antarctica was decided when a New Zealand Jeep was confronted, head-on, by a US Navy Caterpillar tractor weighing 35 ton. Fearing for his life, the New Zealand driver gave way. It occurred to me that most international rules have been settled out of the barrel of a gun. The principle is the same, but in Antarctica we are more civilized.

There were two main kinds of building. One was rectangular and made from prefabricated plywood panels with thermal insulation between. Small double-glazed windows punctuated their orange-painted sides. The length of a building depended on how many panels had been assembled and joined together. The panel buildings were for the upper classes, mostly officers and petty officers, and for administrative offices. I was berthed in one named "The Squirrel Cage." Other buildings, known as Jamesway huts, were salvaged from the

1950–1953 war in Korea. They consisted of an insulated canvas "blanket" spread over plywood arches rising from an insulated plywood floor. They were drab green in color and most of them had no windows. Many junior ranks and some of the USARPS were housed in Jamesways. There was a kerosene-burning stove in the

fully occupied, it was a good place to pick up gossip. The hut was heated, but some well-intentioned electrician had installed an extractor fan in the roof. The result was updrafts of odorous, freezing air directed at the most vulnerable parts of the body. Reversing the fan would have avoided any discomfort and kept us warm. Nobody

McMurdo's "Main Street" in January 1960.

middle, and if you were lucky, you could choose a camp cot either near the stove or far from it. Close to the stove, blankets were not needed, whereas at the ends of the hut we needed three or four to keep warm. Fumes were unavoidable. Showers were rationed to one a week.

The privy, or "head" in naval terms, was a diabolical structure built over a row of empty fuel drums. There were no partitions and thus no privacy, so when

lingered. Perhaps it was by design.

There were other, larger buildings made from corrugated steel arches and in use for communal dining ("the mess hall"), vehicle garages, and supply warehouses. More than 500 people could be temporarily housed at McMurdo during the Antarctic summer. They had to be fed; appetites were voracious. Menus were published at the start of each week. The week beginning 25 January 1960 announced these menus for Tuesday:

Breakfast
   Canned cherries
   Assorted cereals
   Hot farina
   Grilled cheese omelet
   Hash brown potatoes
   Spanish sauce
   Hot toast
   Sweet rolls
   Hot and cold drinks
   Fresh New Zealand milk

Dinner
   Pepper pot soup with crackers
   Stuffed pork chops
   Mashed sweet potatoes
   Chinese fried rice
   Country cream gravy
   Steamed sauerkraut
   Harvard spiced beets
   Spiced apple sauce
   Bread, butter, jam
   Dessert table
   Hot and cold drinks

Supper
   Cream of tomato soup, crackers
   Oven baked beef loaf
   Hash brown potatoes
   Baked macaroni: tomato bacon
   Brown onion gravy
   Buttered brussel sprouts
   Buttered wax beans
   Bread, butter, jam
   Dessert table
   Hot and cold drinks

For those still hungry, there was "midnight chow," ostensibly for night watchkeepers but also patronized by those who found that after three big meals a day, there was space for a fourth. All this was far removed from Bolton rations and more than anything I had ever been offered in the Antarctic. The initial reaction was to overindulge, but gluttony led to discomfort, and after a few days, discomfort led to discretion. When steaks were served, there was a giant oven-pan of rare, another of medium, and a third of well-done meat. The problem was that every steak weighed at least 16 ounces and that was too much for me. My mother had taught me never to leave food on my plate, whereas American mothers long ago gave up trying. The cooks refused to offer half a steak, so half of my steak ended up on the garbage dump — which was bulldozed into the sea. Elsewhere in Antarctica, the Russians kept pigs to recycle their garbage, enjoying fresh roast pork on special occasions.

One of the many things to amaze the visitor was the vehicle "dead-line." Here were rows of experimental tractors, trucks, and oversnow vehicles of many sizes, evidently abandoned for lack of spare parts or because they were considered unsuitable for Antarctic use. One was in mint condition, and when I asked Jack Long to explain, with a twinkle in his eye he suggested "Maybe the ash tray was full!"

The edge of the settlement, particularly towards Scott's *Discovery* expedition hut, resembled a vast rubbish dump. Empty fuel drums, abandoned tractors cannibalized for spare parts, dunnage from ships, and broken building panels lay all about. It was a travesty of civilization, and years later, when Greenpeace came visiting, it gave their people a field day.

The citizens of McMurdo, if we can call them that, came from all parts of the world and all strata of society. Though the official language was English, the common currency was expletive. Paul Siple, a scientist who took part in all of Byrd's Antarctic expeditions between 1929 and 1956, observed that "At McMurdo the use of profanity had been developed into a fine art."[10] I myself served as an Ordinary Seaman in the Royal Navy and had mixed with seamen from a variety of cultures, but my ears were assaulted with obscenities that were new to me. However, the atmosphere was friendly and cooperative — no stigma was attached to speaking without expletives.

The boss of the whole naval "task force" — Operation Deep Freeze — was Rear Admiral David M. Tyree. He was a quiet, kindly, and somewhat shy man. As a USARP with already two years of Antarctic experience, I was considered a cut above the average. The admiral entertained me in his quarters. I told him of my hopes for working in the Transantarctic Mountains.

The question forming in my mind was how to measure the volume of ice that was flowing into the Ross Ice Shelf from the major valley glaciers that transect the mountain range. Many areas at the foot of the mountains were known to be badly crevassed, so I was afraid to use the large tractors favored by USARP. Only two months before, Tom Couzens, a New Zealander, had been killed when his two-ton Sno-cat broke through a snow bridge and fell 100 feet.[11] The research I wanted to undertake would require traveling in the same area. The obvious solution was to use dog teams and light sledges to spread the load in crossing crevasse snow bridges. Dogs, moreover, never suffer from carbu-

The outskirts of McMurdo in 1960. Captain R. F. Scott's *Discovery* expedition hut (right) dates from 1902.

Rear Admiral David M. Tyree.

retor icing or worn sprockets. A dog team was kept at McMurdo and I asked permission to use it. "No," said the admiral; it was only for emergencies, ready to be dropped by parachute to rescue a field party in distress. I would have to think again.

1 R. F. Scott. *The Voyage of the Discovery.* London, Smith Elder & Co., 1905 (2 vols).

2 Roald Amundsen. *The South Pole. An Account of the Norwegian Antarctic Expedition in the "Fram", 1910–1912.* London, John Murray, 1912 (2 vols).

3 E. H. Shackleton. *The Heart of the Antarctic, being the Story of the British Antarctic Expedition 1907–1909.* London, William Heinemann, 1909 (2 vols).

4 Ernest Joyce. *The South Polar Trail. The Log of the Imperial Trans-Antarctic Expedition.* London, Duckworth, 1929.

5 Laurence McKinley Gould. *Cold, the Record of an Antarctic Sledge Journey.* New York, Brewer, Warren & Putnam, 1931.

6 Richard Evelyn Byrd. *Little America. Aerial Exploration of the Antarctic, the Flight to the South Pole.* New York, G. P. Putnam's Sons, 1930.

7 Sir James Clark Ross. *A Voyage of Discovery and Research in the Southern and Antarctic Regions during the Years 1839-43* (2 vols). London, John Murray, 1847 (Vol. 1, p. 228).

8 *Survival in Antarctica.* Washington, DC, National Science Foundation (n. d.), (p. 31).

9 Sir Vivian Fuchs and Sir Edmund Hillary. *The Crossing of Antarctica.* London, Cassell, 1958 (p. 111).

10 Paul Siple. *90° South. The Story of the American South Pole Conquest.* New York, G. P. Putnam's Sons, 1959 (p. 138).

11 *Polar Record,* Vol. 10, No. 66, 1960, p. 280.

Chapter Three

Echoes of the Heroic Age

> *One half of the world cannot understand the pleasures of the other.*
>
> Jane Austen (1775–1817)

Some of my colleagues flew home shortly after we arrived at McMurdo, but I was intent on staying as long as possible to do some local glaciological work and to prepare for what I hoped would be another one or two seasons' work in the area. With this in mind I took every opportunity to go on reconnaissance flights in any direction. Without knowing what the terrain looked like, it would be impossible to select the right equipment for next year. George Toney was supportive.

The naval air contingent was known as Air Development Squadron Six (VX-6 for short). Their fleet at McMurdo consisted of the naval versions of:

- One Douglas C-54 Skymaster (4-engines)
- One Lockheed C-121J (R7V) Super Constellation (4-engines)
- Three Lockheed P2V (LP-2J) Neptune (4-engines)
- Six Douglas R4D (LC-47) Skytrain (2-engines)
- Three Sikorsky HUS-1A (LH-34D) helicopters (single-engine)
- Three De Havilland UC-1 Otters (single-engine)

Some of the aircrews were keen, a few indifferent. The Officer-in-Charge of the wintering contingent was Lieutenant-Commander Robert L. (Bob) Dale, USN. He was a geologist by training and understood what science was about. It was a rare privilege to work with someone who was as keen as the USARPS to grasp every opportunity to study the area and who had the means to get us where we wanted to go. After I had made the case, Bob quite often suggested doing more while we were about it. He even came with us whenever he could. On retiring from the Navy a few years later he was invited to work at NSF and did so.

On 14 January I joined a small group for a day trip to Cape Crozier at the eastern end of Ross Island, the site of an emperor penguin colony made famous by Edward Wilson's winter journey in 1911 and Apsley Cherry-Garrard's account of it.[1] Edward Wilson,[2] Scott's medical officer and chief scientist, led the party together with H. R. Bowers[3] and Cherry-Garrard. They man-hauled two heavy sledges from Cape Evans, on the west side of Ross Island, to Cape Crozier, a distance of 60 miles. Their purpose was to obtain embryos from emperor penguin eggs to study their evolutionary biology. This was believed to be possible only at the coldest time of the year when the eggs were being incubated. Wilson's small party had to endure temperatures as low as -77° F. To protect themselves from extreme winds, they built a shelter with low walls of loose rock roofed with tarpaulin. After suffering extreme hardships, they succeeded in dissecting some penguins, obtained a few eggs, and struggled back to Cape Evans.

Fifty years later, our world was different. Wafted across Ross Island by helicopter, we reached Cape Crozier 30 minutes after leaving McMurdo. The sun was shining and the air temperature was a warm +19° F. Besides the aircrew, the group consisted of Lieutenant-Commander James Lennox-King of the Royal New Zealand Navy, who was to spend the winter as leader of Scott Base, and Lin Gressitt, the entomologist from Hawaii. The helicopter dropped me off at the foot of The Knoll, a small volcanic cone overlooking the ice shelf. Climbing to the summit, I was confronted by a panorama of chaos where the ice shelf is deflected

One of the six R4D aircraft. *Que será será* was the first aircraft ever to land at the South Pole.

McMurdo Ice Shelf (the pinnacled ice), looking north from Black Island.
The giant eddies in the moraine-covered ice have yet to be explained.

and torn asunder by its rapid flow past the cape. Far below, at the foot of a 1,000-foot-high cliff, lay a stretch of fast ice between giant rifts. That must be where the emperors spent their winter. The helo rattled on its way and left me for an hour to take in the scenery.

When it returned, we went searching for the remains of Edward Wilson's rock shelter, soon finding the site on a small and narrow ridge overlooking a steep ice slope leading down to sea-ice. All that remained was a rectangular pattern of rounded blocks reminiscent of an eskimo winter house. The entrance faced north. Wilson's party had assembled the heaviest boulders they could move. We knew from their story that a sledge had been rigged as

Russian exchange scientist Sven Evteev working the generator to power our radio.

a ridge pole to support the canvas roof. The shelter was filled with snow, but digging in it, we found flippers and skin fragments of dissected penguins. Wilson had sacrificed emperors not only in the cause of science but also to provide fuel for a blubber stove. There were fragments of green tarpaulin and remains of a packing case containing smelly bits of manfood. I was surprised that no skua in 50 years had been hungry enough to eat it. South polar skuas *(Catharacta maccormicki)* are the local scavengers, but because of the proximity of penguin colonies, Cape Crozier skuas could afford to be choosy in their diet.

Lennox-King found a variety of artifacts in the wreckage, made a list, and carried everything back to Scott Base. Many are now on display in the Christchurch Museum. Close by we found a rock cairn containing boxes of modern sledging rations and two jerrycans. We guessed that it had been left by Ed Hillary, who in the course of training his party for travel, led a tractor traverse to this spot from Scott Base in March 1957. His companions were Jim Bates, Murray Ellis, and Peter Mulgrew.[4]

Two hours after returning to McMurdo, I hitched a ride on a "bug run." Bug runs were trips in an Otter flying 10–

30 feet off the ground. Lin Gressitt was hunting for airborne insects. Few of us see Antarctica as an obvious haven for insects of any kind, but Gressitt was a man with a mission. An entomologist with a butterfly net could run for miles in these parts without catching a single insect. So in order to stand a better chance, he hit on the idea of flying the nets. Both cabin doors of the aircraft were removed and in their place was rigged a pulley arrangement to deploy nets on both sides of the aircraft. They were aerial plankton nets. Describing them as butterfly nets was frowned on as being frivolous. The doorless cabin was breezy for passengers but we thought it a fair price to pay for seeing the countryside.

The pilot was Lieutenant Donald (Don) Moxley, USN. He flew across McMurdo Sound at a height of 10 feet over the sea — a good test of pilot concentration. Increasing the safety margin to 30 feet over the land, we crossed Wilson Piedmont Glacier and glided down into Wright Valley, a snow-free desert of a place that could be mistaken for the Sahara. We saw rolling sand dunes and every sign of dessication. Yet small streams from glaciers 30 miles apart at either end of the valley flow into Lake Vanda. Vanda has no outlet; evaporation (or sublimation) is the only thing that prevents the whole valley filling with water — or ice.

Climbing through Bull Pass we came to Lake Vida in Victoria Valley, another vast desert with no outlet to the sea. Prospective astronauts have been taken into the dry valleys because the surface of the Moon is not so very different. After 2½ hours we landed at McMurdo, chilled to the bone but thrilled to have had a guided tour of the dry valleys about which I had read so much. Gressitt wound in the nets: both were empty.

Still determined, he later arranged for a net to be deployed from the Super Constellation that carried passengers between Christchurch and McMurdo. Surely, he thought, flights over a distance of 2,000 miles must trawl in some airborne insects. But no. One day the pilots played a practical joke by opening the cockpit window and throwing some dead flies into the net. Gressitt was not amused — or fooled.

I soon got used to days of waiting between flights. This was either due to bad weather or to the fact that scientists pursuing many other projects needed to use the same aircraft. But summer weather at McMurdo is sometimes good for long periods, with temperatures in the range between 0° F and 30° F.

On 15 January I arranged an Otter flight 200 miles south along the Transantarctic Mountains to reconnoiter the surface of some of the big eastward-flowing glaciers. For the following year, I needed to know whether they could be safely traversed for setting out ice movement markers, or whether I would have to seek helicopter assistance. Don Moxley was again the pilot. Heading south between White and Black islands, we climbed over Minna Bluff, a long arm of rock and ice stretching eastwards from the 9,000-foot ice-capped volcano Mount Discovery. The first big glacier was Skelton. We knew it was reasonably safe for ground travel because Sir Vivian Fuchs' Trans-Antarctic Expedition used it in 1958 to descend from the high plateau down to the ice shelf.[5] They had some

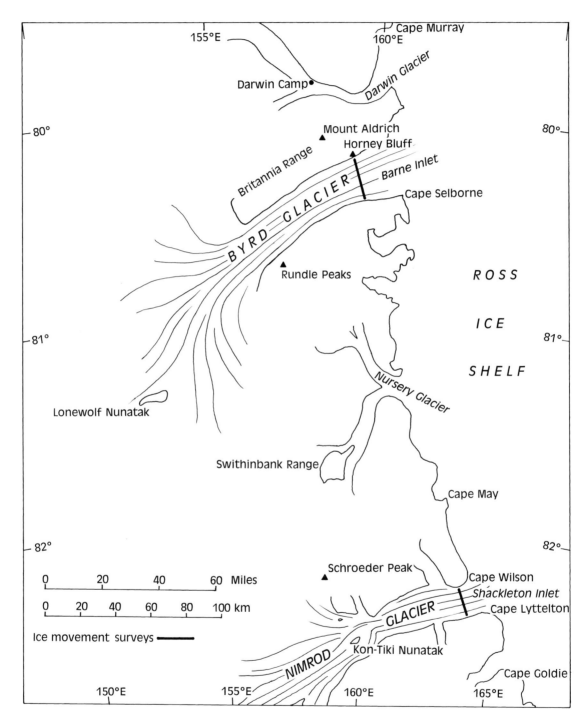

Byrd Glacier to Nimrod Glacier.

Byrd Glacier (looking upstream) from an altitude of 9,000 feet. This glacier discharges more ice than any other in the 2,000-mile-long Transantarctic Mountains. It is 16 miles wide where we set out our stake line. The surface is so badly crevassed that I had to be winched down to it from a helicopter.

surprises with crevasses higher up, but I was intent on working at the lower end.

From here southwards, the existing sketch maps did not show which glaciers flowed from the high plateau and which took their nourishment largely from snows that fell among the mountains. I needed to know because I would not have time to study them all, only those that carried ice from the plateau. Ed Hillary's surveyors on the Trans-Antarctic Expedition had done some fine surveys but their maps were not yet published. The next glacier was mapped only as Mulock Inlet, named after Lieutenant George Mulock of the Royal Navy, a surveyor with Captain Scott's first expedition of 1901–1904. The glacier was eight miles wide. Not only were the sides inaccessible from the glacier surface but it was uncrossable because of uninterrupted crevassing from wall to wall.

The third big glacier also had no name. In 1902, Scott's southern party had sledged past it at some distance but only saw it as an inlet in the mountain chain. A year later it was approached by a survey party led by Lieutenant Michael Barne, RN, so became known as Barne Inlet. It took only a glance at the chaotic surface of the ice to understand that there was no way in which survey markers could be deployed here without helicopters. So we returned to McMurdo after 4½ hours in the air.

Three days later I was hitch-hiking on another bug run. There were two other passengers and both of them slept throughout the flight. Asking why they had come, it was explained that if sailors flew a certain minimum number of hours per month, they qualified for "flying pay." With Don Moxley again as pilot, we flew to the Dailey Islands, Cape Chocolate, Koettlitz Glacier, and Black Island, keeping all the time 10–30 feet above the surface except where rough terrain forced us higher. The area is exceptional because more ice melts off the surface in summer than can be replaced by snow in the winter, whereas over most of Antarctica snow accumulates at all seasons. The result is a spectacular landscape of dazzling ice pinnacles separated by torrents of melt water. Some of the pinnacles were 60 feet high. In places we could peer into the deep blue of lakes beneath clear ice. The calm serenity of the scene looked like the stuff of fairy-tales, and I resolved to return to take a closer look.

On 24 January I hitched a ride on a geological reconnaissance flight over the dry valleys. Dickie Wirdnam was pilot. John Mulligan, a geologist with the Bureau of Mines, proposed to look at the terrain to see what kinds of rock were exposed and whether it would be easy to travel the area on the ground. For a geologist, this was a nostalgic excursion over the routes of the British geologists who explored the area from 1902 to 1912. We recognized the place names as being those given by these earlier explorers. Flying across McMurdo Sound to Gneiss Point, we again crossed Wilson Piedmont Glacier, the long coastal strip of ice that drains into the dry valleys on one side and into the sea on the other side, and on over Debenham Glacier, named after a geologist on Scott's 1910–1913 expedition who later became my friend and mentor.

Climbing laboriously to 10,000 feet to avoid some mountain turbulence, we had a graphic demonstration of why the dry valleys are snow-free. A blizzard of snow was flying off the high plateau to the west and over the dry valleys, where it began to fall. But as fast as it fell, the snow turned into vapor. The process is known as sublimation because the snow vaporizes without melting. Evidently the warmer dry air at lower levels was able to take up all the moisture precipitated into it from above. Scientists have published elaborate theories about why the dry valleys are dry, but here the explanation was revealing itself before our eyes.

On the way home we had a grandstand view of the Royal Society Range, whose peaks rise to 13,000 feet and dominate the view west from McMurdo Sound. The lasting memory of this flight was savoring the contrast between Debenham's laborious progress in 1912 and our passage over the same landscape at 120 knots in a heated cabin. We felt humbled.

Five days later I was corralled to fly with an R4D carrying fuel up Ferrar Glacier to the plateau. The glacier was named after Hartley Ferrar, a geologist with Scott's first expedition of 1901–1904. Bob Dale was pilot with Ron Carlson as copilot. This flight was not sightseeing, but instead, sweated labor. We were carrying drums of fuel to make a cache for an inland tractor traverse and the crew needed muscle power to unload and stack the drums. Routing us close along cliffs of alternating black and brown strata, Bob did not hide his interest in the geology of the area. Some Navy pilots thought of themselves as mere taxi-drivers for the scientists, so it was refreshing to fly with a pilot who saw himself as a colleague. On the ground in a temperature of -4° F and at an altitude of 7,000 feet barely an hour after leaving sea level, we were quickly exhausted with the effort of rolling the drums through soft snow. Some of us found ourselves coughing up blood, a classic indicator of frostbitten air passages. It is not healthy, but neither is it critical for fit people. Luckily there were no after-effects.

On 2 February I got to fly in another R4D southwards along the Transantarctic Mountains. Ron Carlson was pilot and the VX-6 squadron commander, Captain William (Bill) Munson, USN, was copilot. We took off with JATO. The letters stand for Jet Assisted Take-Off but in practice it means rocket-assisted take-off. Four solid fuel rockets about the size and shape of a 100-pound bomb were mounted on each side of the fuselage. When a critical speed was reached, the pilot threw a switch.

Twenty years later someone might have exclaimed, "We have ignition!" But these were early days for spacecraft, and whatever anyone did say was drowned by a deafening roar close behind our seats. The noise was so loud that first-timers could believe that their lives were about to terminate in a fireball. A common practical joke was to seat newcomers just where, through the thin aluminum skin, the rockets were mounted. Old timers looked on with glee as naked terror swept over the face of the new boys. I was the unfortunate new boy and it took my breath away.

On the way we enjoyed glorious panoramas of ice-clad, glistening peaks, some of them reaching to a height of 14,000 feet. We landed at Beardmore, a small summer-only weather station 400 miles from McMurdo and 30 miles off the mouth of Beardmore Glacier. It was the only inhabited spot in 200,000 square miles of the Ross Ice Shelf and consisted of one Jamesway hut. The total population was three enlisted men whose duty was to radio weather reports to McMurdo every three hours. The main purpose of the flight was to carry fuel drums. Our other task was to determine the new position of a snow stake that Bert Crary had erected two years earlier after fixing its position relative to mountain peaks. Nobody had any idea of the rate of ice movement in the area and he (and I) needed to know. John Mulligan was my helper for the day. Together we searched with binoculars and found the stake a mile away. By taking angles to the same peaks that Crary used, we could establish its position with some precision. It had moved at a rate of 1,214 feet per year towards the north-northwest, indicating that ice from further east along the mountain range had already deflected to the left the flow coming from Beardmore Glacier.

On the way home we flew close to the mountains. I had once written Jim Zumberge that "A trip to the foot of the Beardmore . . . would be like going to Heaven or near it." Now I was there. The great white ribbon of ice coursing towards us was fourteen miles wide at its mouth; it seemed to be forcing the mountains apart. At its foot lay Mount Hope, from where on 4 December 1909 Shackleton first saw this giant glacier that gave him the prospect of a road towards the South Pole.[6] We marvelled at his luck in finding a path through tortuous crevasse fields at the junction between floating ice and grounded ice at the foot of the mountain range. Passing Cape Wilson, we saw the place where, on 30 December 1902, Scott, Shackleton, and Wilson had turned homewards from their farthest south.[7] They were stopped by disturbed ice and giant rifts where the 11-mile-wide Nimrod Glacier flows into the ice shelf. Nimrod was one of the glaciers that I proposed to work on the following year, so I tried to memorize where the safest areas were.

On 4 February, I was able to camp in McMurdo Sound on the "pinnacled ice," as Frank Debenham had called it. Joined to the main part of the Ross Ice Shelf, the ice is relatively thin and its surface is only about 10 feet above sea level. Narrow rivers of melt-water flowing during the day often made it necessary for us to travel over the ice by night, when the streams dried up. With me was a 25-year-old Russian exchange scientist, Sveneld Evteev, who like me was a glaciologist. He was due to winter at McMurdo. Never having been outside the Soviet Union, he looked bewildered at finding himself in what was, to all appearances, a US Navy base. Schooled in the Cold War, I could see that he had difficulty reconciling all that he had been taught about Americans with the easy-going, friendly people he was now living with. He showed little initiative and seemed to have no particular projects to pursue.[8] We could only presume that he was used to being told what to do whereas we more often told our superiors what we proposed to do. Lin Gressitt came along in order to empty some nets that he had set out earlier in the season.

I wanted to set up three markers on the ice shelf between Mount Discovery and Black Island and then to determine their position relative to fixed points on rock. By repeating the measurements the following year, we would calculate how far the ice had moved in the interval. Krebs carried us by helicopter to a flat patch of moraine at the foot of Mount Discovery. We had camping gear, food, and fuel for a couple of weeks, even though the work was planned to take no more than three days. On the 1949–1952 expedition I had learned the hard way that being hungry was no fun when one realised that it might continue for an indefinite period. Having set up our stakes and taken angles to them, we were picked up two days later. My diary for the day says "I have a miserable cold, my first ever in the Antarctic. Civilization is catching up with us."

Byrd Glacier seen from the Landsat-1 spacecraft orbiting at an altitude of 570 miles. The full width of the image covers 110 miles on the ground. Byrd Glacier flows from the bottom of the picture and discharges into the Ross Ice Shelf at the top. Darwin Glacier is to the left of Byrd Glacier. Mulock Glacier is at top left.

On 9 February, Evteev and I were taken by helicopter to Cape Crozier to erect markers for measuring ice movement. We unloaded our camping gear and considered how to go about the task. The camp was separated from the proposed position of the markers by a broad belt of severe crevassing on the ice shelf. Without the helicopter, we might have reached the target after a couple of days of hazardous climbing. With it, we were there in minutes, the pilot waiting while we drilled holes and set up three bamboo poles on the ice shelf at intervals of three miles in a southeasterly line from the cape.

Back in camp, a welcome silence followed the departure of the helicopter. Sven and I were alone in a tiny mountain tent perched on a rocky plateau. Incongruously, the tent was provided with a zipped mosquito net across the entrance. We fell to wondering who had ordered our equipment and on which distant continent the nearest mosquito might be found. After a snack, we measured a 1,700-foot-long baseline with a steel tape. Screwed to a stable tripod, our theodolite was capable of determining the angle between any two points sighted through the telescope to better than one millionth of a 360-degree circle. However, to achieve this precision, a surveyor has to be well-trained and extremely patient.

With freezing fingers and eyes made watery by the wind, it was not easy. We searched through the telescope for the distant markers and homed in on the tiny, shimmering, inverted image of each bamboo pole. I read the angles from the instrument's horizontal and vertical scales while Sven wrote them down.

We sent a radio message to McMurdo reporting that our work was done, but it was five days before anyone came to get us. The radio that we had been given was a far cry from what people use today. Sven stayed outside the tent and had to pedal furiously on a hand generator to provide the power while I sat in the warmth inside with a Morse key. Sven's task was relatively easy while I was just listening, but transmitting was another matter. The drag on the generator was exhausting, and Sven knew that if he slacked off, the message would not get through and McMurdo would ask us to repeat it. Each day we told them what the weather was doing but they were busy with other jobs.

Cape Crozier is known not only for its emperor penguins but also for its colony of about 100,000 breeding pairs of Adélie penguins. Having time to spare, we could go and see it. Neither of us knew where to look, but when the wind blew from the north, we were left in no doubt. The stench was dreadful. Hiking upwind, we soon came upon them. Here there was no cliff, so the penguins could waddle in and out of the sea across a beach. There was a constant stream of them both coming and going, each adult intent on returning with enough krill in its crop to provide for the voracious appetite of a fast-growing chick, or sometimes a pair. Some of the chicks were scruffy, muddy, fat, nearly as big as their parents, and busy molting gray-blue feathers to reveal their pure white front beneath. Soon they would be off to sea, fending for themselves, a few unwary ones destined for the jaws of a leopard seal or killer whale.

Having spent a month living with foul smells on the whaling factory ship in 1949, I was able to cope with the stench. But Sven was sensitive and lost his breakfast. We returned to the tent, to our sleeping bags, and to paperback books to pass the time. The days were punctuated only by meals consisting of the monotonous Bolton rations cooked over a Coleman stove. When we finally heard the helo, there was a mad rush to pack up and load, because most pilots were afraid to shut down the engine in case it failed to restart.

On 26 January, seven Lockheed LC-130 "Hercules" aircraft belonging to the US Air Force arrived at McMurdo from New Zealand. These are four-engined military transports, the largest aircraft ever put on skis. They came at the request of the Navy because the Navy was interested in purchasing the same type to supply the South Pole and other inland stations.

For my own work, there were more helicopter flights for ice movement studies. I was humbled to think of the cost of all that VX-6 had done for me. But we had achieved a lot in a short time, so had no regrets.

Jack Long and Sven Evteev left McMurdo with two companions on 25 February driving two giant Sno-cats to do some seismic ice depth sounding work south of Minna Bluff. It was the first test of Sno-cats specially built to the largest size that could be driven into a Hercules with just inches to spare. Each vehicle weighed 10 ton. They began to break down within a few miles, so Jack had to call for spare parts to be airdropped. By the third airdrop the pilots were becoming mildly irritated, so one of them dropped leaflets. Addressed to the "Mañana Traverse Party" a drawing showed one of the men calling "Where's my mother?" and another requesting some lollipops inadvertently left behind. It was some time before there were more calls for help.

At this stage in the season the weather

USS *Atka,* the last ship out of McMurdo, 13 March 1960.

was beginning to get chilly, hovering around -5° F, the temperature of a meat-freezer back home. So I bade farewell to many new-found friends and arranged passage northwards in the icebreaker USS *Atka*, the last ship to leave McMurdo that season. *Atka* was one of seven *Wind*-class icebreakers. Four were built during World War II and named after the cardinal points of the compass. When more of the Wind class were needed, the authorities faced a naming dilemma. Rather than launching a Northeastwind, and so on, they resorted to other names. Like some fickle woman, *Atka* had changed her name three times. She began life in 1944 as the Coast Guard cutter *Southwind*. Loaned to the Soviet Union in 1945, she took the name *Kapitan Belousov*. Returned to the US in 1950, she became USS *Atka*. But only weeks after I sailed in her she came full circle, returning to the Coast Guard and claiming back her maiden name of *Southwind*.

Icebreakers can be described as very heavily reinforced ships with engines yielding more than about one horsepower per ton of displacement. *Atka* had 1.5-horsepower-per-ton compared with *Arneb*'s 0.4 horsepower-per-ton. On 13 March *Atka* was frozen in at McMurdo and we could walk round her, but as soon as the throttles were opened she was breaking ice. In open water some hours later, it took eight days of pitching and rolling through equinoctial gales to reach Lyttelton.

Once more in Christchurch, half an hour's drive from Lyttelton, there was the joy of seeing green grass and feeling breezes blow through lightweight summer clothing. Civilization has its attractions.

---

[1] Apsley Cherry-Garrard. *The Worst Journey in the World.* London, Chatto and Windus, 1922 (2 vols).

[2] George Seaver. *Edward Wilson: Nature-lover.* London, John Murray, 1937 (2 vols).

[3] George Seaver. *'Birdie' Bowers of the Antarctic.* London, John Murray, 1938.

[4] Mulgrew lost his life in November 1979 when a DC-10 aircraft on a sightseeing flight from New Zealand crashed into Mount Erebus.

[5] Sir Vivian Fuchs and Sir Edmund Hillary. *The Crossing of Antarctica.* London, Cassell, 1958, (pp. 289–291).

[6] E. H. Shackleton. *The Heart of the Antarctic, being the Story of the British Antarctic Expedition 1907–1909.* London, William Heinemann, 1909 (2 vols).

[7] R. F. Scott. *The Voyage of the Discovery.* London, Smith Elder & Co, 1905 (2 vols).

[8] Some years later, Sveneld Evteev became Assistant Executive Director of the United Nations Environment Programme, based in Nairobi, Kenya.

Chapter Four

Into the Unknown

It is quite impossible to speak too highly of my companions.
R. F. Scott, 8 January 1912

People sometimes ask: "What do you do when you are not on expeditions?" The question implies that life between expeditions must be tediously uneventful. But reality is different. Four months a year in the Antarctic leaves only eight months in which to write reports to sponsors, search for and recruit people for the next field season, discuss what equipment to take, and then order a thousand items from suppliers all over the world, each item to be delivered without fail in time for a shipping deadline in September. In the case of critical instruments or supplies, a single late delivery can lead to cancellation of the whole field season. Some manufacturers seem not to understand the concept of an absolute deadline. So as September approaches, the telephone is in constant use to charm, cajole, or threaten late suppliers.

All these things are urgent enough. However, there is one thing that cannot be postponed, and that is the task of seeking funding for the season after next. Under the federal system of managing Antarctic research, this year we are living off the results of last years's carefully-argued proposal that extolled the merits of our case. There is no guarantee that funding will continue from one year to the next. The proposal must have signatures, not only from the leading scientist involved but also from senior officials of the university agreeing to house and administer the research grant — if it ever comes. NSF in Washington, DC, sends the proposal for "peer review" to a dozen or so university or government scientists. Each should be an expert in the general field covered by the proposal. They are asked to judge the merits of the case. Peers are sometimes one's best friends, but equally, they may be competitors who would prefer to pursue the subject themselves without competition. However, rather than compromise their integrity, they generally give pause on considering that next time, the tables could be turned — all of us review proposals. Although the process is confidential, some reviewers choose to identify themselves and may correspond with the proposers to suggest improvements.

With only an off-the-record nod from

NSF on the prospect of this year's funding, I launched into the summer routine. The first task was to make up a team. A party of four would provide a reasonable margin of safety, so I pored over the personnel files of aspiring explorers. The first was easy: David C. Darby, a 28-year-old paleontologist at the University of Michigan, was also an inveterate traveler who longed for new adventures. Small in stature, he was tough as nails yet easy to get along with. He had married two years earlier and for a honeymoon took his bride to hunt the abominable snowman in the Himalayas. Getting within 15 miles of the territory in which this creature is supposed to live, he was turned back at the border with Sikkim. As neither he nor his wife were Buddhists, the border officials would not let them in. Whereupon the couple went on to Angkor Wat in Cambodia before making a night climb of the Great Pyramid in Egypt.

Then there was Jack Tuck, a tall, handsome, bearded 27-year-old who in 1957, as Lieutenant (j.g) John Tuck, Jr., USN, had commanded the naval support unit at the Amundsen-Scott South Pole Station during its first winter. A graduate of Dartmouth College, Tuck first wintered at McMurdo Station in 1956. As a naval reserve officer in New Hampshire, he had learned to train and drive sledge dogs and spent his first year in the Antarctic as officer-in-charge of 31 huskies. In the spring of 1956, eight men and two dog teams were landed on the polar plateau near the South Pole.[1] Tuck and three

Adélie penguins (*Pygoscelis adeliae*) discussing the weather.

others drove the last eight miles to become the first Americans to reach the pole in the old-fashioned way. Among other attributes, he was the man to guide us through the maze of officials at McMurdo.

I needed a surveyor, not so much to navigate in the wilderness as to make maps. The only maps of where we were going were the sketches made by members of the Scott, Shackleton, Amundsen, and Byrd expeditions. I asked the US Geological Survey (USGS) if they could come up with the right sort of man. They found Thomas E. Taylor, an experienced field surveyor with a reputation for backpacking theodolites up mountains with the greatest of ease. He had been attracted to USGS as a career, he said, because it meant "being paid to climb mountains." Aged 37, Tom was the old man of the party, four years my senior. He hailed from Telluride, Colorado, and had served in the Navy both in World War II and in the Korean War. He had earned B. S. and M. S. degrees in Mechanical Engineering, later adding Applied Mathematics to his armory. He had practiced and also taught geodetic survey, so came with the skills that we most needed.

All three of my volunteers had full-time jobs at home and would be available only for the field season. None was in a position to share the preparations. Our biggest problem was to be the crevassed nature of the terrain where the big glaciers flowed through the Transantarctic Mountains. This 2,000-mile-long chain of mountains contains some of the most spectacular glacial scenery to be found anywhere on Earth, but also some of the greatest hazards for the traveler. Every one of the glaciers is heavily crevassed where it cascades into the ice shelf. No party in the history of Antarctic science had planned to work in areas potentially so dangerous.

Seeing myself as a risk manager rather than a risk taker, I decided to try new methods of reaching the glaciers. Helicopters were the obvious answer. But only the first two glaciers were within helo range of McMurdo Station. The rest had to be approached over the surface, in places riddled with crevasses. Open crevasses are not inherently dangerous because you can see them and travel along them until you get to the end. Snow-bridged crevasses, on the other hand, are sometimes invisible. Visible or not, there are millions of them, and you have to cross them. Snow bridges can be thick enough to support a tractor or, at other times, thin enough to give way under the weight of a man on skis. The only way to find out is to dig or probe through the bridge to establish its thickness. But doing that, we would get nowhere because of the time involved.

We could not avoid risk but we aimed to reduce it to an acceptable level. It was no use being heroic if only empty-handed survivors were to come home. Falling down an average crevasse can be compared with falling from the roof of a 10-story building. Unhealthy — as we used to say. Teams of dogs pulling sledges provide the classic solution. Loads are spread out, and although a few dogs may fall in, the sledge stays on the surface. The dogs are then pulled out by their harnesses. But the dog solution was not available — by order of Admiral Tyree.

Fred Roots, a Canadian geologist who had been a fellow member of the Norwegian-British-Swedish expedition, suggested motor toboggans. These were, in principle, similar to the vehicles that were taken south earlier in the century by Shackleton and Scott. After the 1960s they came to be known as snowmobiles.

Nobody had used anything like them in the Antarctic since 1911. But we had we asked for a generous set of spares. Too often in Antarctica, when a machine breaks down hundreds of miles from anywhere, we unpack the manufacturer's instruction manual, only to find these words: "If you have any problems with your purchase, we recommend that you

Double-walled Scott (Pyramid) tents kept two of us snug in any kind of weather.

no choice. So I ordered two Eliason K-12 motor toboggans from the FWD Corporation of Kitchener, Ontario, Canada. The Eliason had a nine-horsepower four-cycle engine driving a single flexible track between two long skis, with two short, steerable skis at the front. The engine was a single-cylinder Briggs and Stratton of the type more often seen on lawn mowers than in the Antarctic. After consulting Harold Serson in Canada about which components were potentially unreliable, consult your friendly neighbourhood dealer." He is invariably 10,000 miles away.

Sledges were needed to tow behind the machines. So I ordered four from Norway, each 13 feet long and 27 inches wide, made from hickory and built according to the classic design of the Norwegian explorer Fridtjof Nansen (1861–1930).[2] Some concessions were made to modern materials, for example plastic laminate glued to the runners to reduce friction. No nails or screws were used. Joints were lashed with rawhide thongs to allow the sledge to bend under load. The design was essentially unchanged from the sledges described by Alan Reece,[3] though we saw no need to follow the British convention of fitting handlebars nor the American convention of using a "G-pole." Each sledge weighed about 100 pounds.

Tents of the (also classic) double-walled pyramid type had to be obtained from England; no manufacturer in the whole of North America had experience of this design. None of the tents available at McMurdo was even remotely suitable for long field seasons far from base. My prime consideration was comfort and warmth for the occupants, together with the ability to ride out hurricane winds safely. Pyramid tents, known at McMurdo as Scott tents, are made in two sizes, one for two people and a larger for three.

For restful nights, we each had a US Army "Bag, Sleeping, Arctic." There was an inner and an outer layer, both feather-filled. The bags could be zipped up to leave the head covered and only the face exposed — hence they were known as "mummy" bags. They were more than adequate for summer travel. There were periods when we only needed one layer, and on the warmest days we lay on top of the bags to keep from over-heating. We introduced three-inch-thick foam mattresses to spread our sleeping bags on because I had learned that other types of mattresses tend to accumulate body moisture.

Apart from the unsatisfactory Bolton rations that we had lived with the year before, USARP seemed to have no experience in designing a nourishing yet lightweight diet. As with dog sledging, our range of travel would depend on the weight we had to carry. So allowing for three months' travel with a party of four, I ordered a ton or so of food from various suppliers in the US and New Zealand.

When finally everything was ready for shipment, the packing list filled 10 pages of single-spaced type. Each of us had to go to the Great Lakes Naval Hospital for a rigorous physical and "neuropsychiatric" examination. Queuing for blood tests and having every visible — and some invisible — parts of the body inspected took all day. An hour with a psychologist and a second hour with a psychiatrist was intended to identify any hidden personality traits that might surface under stress or from the effects of isolation. After grilling me for 55 minutes, the psychiatrist finished his list of routine questions, relaxed and said, "Well, what do you do for a living?" I replied that I worked in the Antarctic. "Oh my God," he exclaimed, "Why didn't you say so?" After a pause, "Tell me — what am I supposed to be looking for in these people?" After a muffled laugh, my answer was to the effect that any normal person who wants to go south will adapt to the circumstances. All the psychiatrist must do is look for signs of mental instability; there is no place for misfits where I was going.

Another USARP institution was a three-day gathering of all field personnel for the coming season at Skyland Lodge in Shenandoah National Park. It was an occasion to get to know the members of one's team — often for the first time — and to be briefed on how to behave at McMurdo and in the field. Harry Francis, a keen climber as well as an NSF official, led us to a neighboring cliff to practice rope-work and crevasses rescue techniques. It was a particular privilege at this time to meet Paul Siple and others who had served in Admiral Byrd's expeditions from 1929 onwards.

In September, Darby and I decided to inspect our mechanical dog teams at the FWD factory two hundred miles away. Only five days before this I had become the proud owner of a pilot's license. So we went to the local airport and rented a Cessna-172. The FWD factory manager was impressed with the apparent opulence of his customers; we did not disillusion him. The Eliason machines appeared to consist of sections of angle iron bolted together. Unsophisticated was the most one could say for them. However, there were advantages in that: a welded frame would have been more difficult to repair in the field. On the way home from Canada we had to land at the first US airport to clear Customs. The manifest and declaration form thrust into my hands ended with a space for signature by the "Pilot-in-Command." It took me a long moment to realize who that was.

We left Ann Arbor on 8 October to fly with a US Navy plane from Andrews Air Force Base to Christchurch, New Zealand. The flight took a week, a far cry from what people are used to in the jet age. We stopped to rest the aircrew and refuel at San Francisco, Honolulu, Canton Island — a tiny atoll — and Fiji. At Christchurch, Eddie Goodale broke the news that flights south were delayed, so Darby and I hired a car and toured the spectacular Mount Cook area.

On 23 October 1960 we boarded the Super Constellation belonging to VX-6 and reached the McMurdo area after 10 hours in the air. However, visibility on the ground was little more than 100 yards and we were in cloud. The pilot made three instrument approaches, each time failing to glimpse the runway. We landed on the fourth attempt. By that time, one engine was backfiring and there were 65 passengers pretending to be unconcerned. Anywhere else on Earth, the pilot would have diverted to an airport with better weather. But we were far beyond the point of no return to New Zealand. Come what may, in Antarctica all but ski-equipped aircraft must put down on the only airstrip there was.

Stepping onto the ice this early in the Antarctic spring, we were grateful to have been issued with a kit-bag filled with a generous supply of cold-weather clothing. In earlier generations, clothing was the subject of endless debate and individual modification with sewing machine or palm and needle. But now it was so good that we seldom discussed its merits. There was something for all tastes, and with care, it could last for years.

A week after arriving, the first-ever Navy C-130BL Hercules landed at McMurdo. The Navy had been so impressed with the performance of the fleet of Air Force Hercules brought to McMurdo in February that they had purchased four of them.

The same day I saw how my landing at McMurdo could have ended. The Super Constellation *El Paisano* landed in bad weather, slewed off the runway, and came to rest minus one wing and with its fuselage broken in half. All twenty men aboard survived, though two were badly injured. The aircraft, a $6 million flying laboratory equipped for geophysical surveys, was beyond repair, although much of the scientific gear was salvaged.

Our first month was spent helicoptering around the local area to remeasure the ice movement markers that had been set out in various places eight months before.

the season before, its transmitter output was only five Watts, so the Morse key was still the normal way to communicate. Although we called in daily exactly on schedule, there were periods when we were out-of-touch for up to a week.

We revisited the pinnacled ice. Walk-

The Super Constellation *El Paisano* after its unhappy landing.

The fastest-moving ice was flowing more than 2,000 feet-per-year, the slowest just three feet. Day temperatures this early in the season were often around 0° F, making long hours of observations at the theodolite less than agreeable.

We had been given a new battery-powered field radio specially designed by a New Zealand company. While this was better than the agonizing hand-generator of

ing near our camp at the Dailey Islands, we were amazed to come across more than 100 dead fish lying on the ice. One was six feet long and weighed over 40 pounds.[4] Nearby there were siliceous (glassy) sponges still attached to pieces of rock. The hairs on the sponges were so delicate that they broke when touched. It became clear that whatever mechanism had brought them to this spot, it must have

David Darby beside a fish that we found on the pinnacled ice.

been inordinately gentle.

In a 1920 journal article, well before our time, Frank Debenham had sought to explain why in this area the ice surface was littered with pebble-sized pieces of volcanic ash and seabed fauna — sponges, corals, and in a few places, even fish. Some of the fish were headless. He surmised that the fish had died, perhaps at the hands (or jaws) of a seal, and floated upwards to be incorporated in the ice with sponges and corals.[5] This developed into a theory of how the ice shelf formed.[6] He proposed that after freezing on to the bottom of the ice the fish eventually worked their way up owing to high rates of surface melting. He had noted that ice crystals form on sponges growing on the seabed and guessed that the buoyancy of the ice tore the sponges away, whereupon they floated to the underside of the ice shelf and became incorporated in it by newly-formed ice crystals. Freezing from below had to balance the rate of melting on top, so that sooner or later, everything freezing on beneath the ice emerged at the surface. The implication that the ice shelf was nourished by the freezing of sea water contradicted the situation elsewhere, where compacted snow progressively turns to ice as it is buried by more snow. Controversial at the time, Debenham's concept was later proved to be essentially correct.[7]

The main season's work did not begin until 22 November. The first sizeable glacier south of McMurdo was Skelton Glacier, but there was no need to study it because Crary had already measured not only its rate of movement but also the depth of ice. His calculations showed that the glacier contributed annually 0.2 cubic miles of ice to the Ross Ice Shelf.[8] At the time, this seemed an impressive volume of ice, but our own studies further down the mountain range were in due course to reveal that it was quite small.

Our first port of call, Mulock Inlet — later known as Mulock Glacier — was within helicopter range of McMurdo, but the loads were strictly limited by weight. The helos were Sikorsky machines of a shape somehow reminiscent of a Trojan horse. Although each could lift more than a ton, on these trips part of their load was taken up with extra fuel. Weighing our equipment, we found that it weighed 2,230 pounds, which was 1,000 pounds more than had been estimated. The pilots, Buddy Krebs and John Hickey, were — to say the least — displeased. Any temptation that we might have had to conceal the overweight vanished with the discovery that we ourselves were to fly with the cargo. Hickey and I flew ahead with the first 1,100 pounds to scout out the land. We chose a camp site near one of the few accessible rock outcrops at the top of the cliffs and unloaded the cargo. We were back at McMurdo, more than 100 miles away, in time for lunch.

The afternoon departure was with two helicopters and an Otter carrying extra fuel for them. The Otter went ahead and landed on a smooth patch of snow off the glacier. Now followed the rather hair-raising events with which I began my story — the high wire act.

Lt. John Hickey, USN, helicopter pilot.

Besides Darby and the aircrew, there were some less welcome witnesses. Whereas most British expeditions treat representatives of the press like lepers, many Americans not only welcome the press but go out of their way to treat them like VIPs, or DVs (Distinguished Visitors) as they were known locally. Wadsworth Likely of the *New York Herald Tribune* was circling us at close range in the second helo, adding to the confusion of noise and

The author expounding to the press.

the difficulty of ensuring unambiguous signals between those on and near the ground.

Back in camp, Taylor had been following our progress through the telescope of a theodolite, taking angles to every stake. But for this, we should never have found the stakes again. The helos headed for McMurdo and a welcome silence reigned. We feasted on the view from the cliff-top camp — the glacier in a trench far below, beyond it the towering ramparts of the Worcester Range rising to 9,000 feet, and to our right the vastness of the Ross Ice Shelf. We spent a week measuring two survey baselines and fixing the position of every one of the stakes with the theodolite. Taylor took sun observations and intersected all the peaks in the area for topographic maps to be made by the USGS. We learned something of the climate by measuring ice temperatures in a drill hole to a depth of 33 feet. At this depth, seasonal fluctuations of temperature at the surface are virtually damped out, giving a good approximation of the mean annual air temperature. Here it was -11° F. We searched for mites and lichens, and Darby pored over the rocks.

Our pyramid tents were designed for

two men. With aluminum poles at each corner, the canvas could be stretched tight and was stable in high winds. The outer skin has a 24-inch snow-flap all round on which snow blocks are piled to hold the tent down. The peak of the tent is seven feet high. An inner skin is suspended from the outer and has an inward flap which, with sleeping bags spread out, prevents drafts. The inside floor area is 87 by 78 inches which allows plenty of room for two sleeping bags with a food box and a cook box aligned in the space between. The cook box contains utensils and matches. Whereas Europeans generally use the tried and tested Swedish Primus stove fuelled by kerosene, Americans prefer a Coleman stove fuelled by unleaded gasoline. Both work well, though clumsiness with either can set the tent on fire. The fumes are vented through a piece of radiator hose passing through both layers of the tent.

Mr. Likely asked if he could share a tent with me for a few days. He enjoyed the respite and spent the time writing notes. Little did I know that some weeks later he was to publish three long articles about us in the *New York Herald Tribune*.[9]

We had not chosen the easiest way to measure glacier movement, though it was the most accurate method available. Aerial photogrammetry — measurement on aerial photographs — has been used successfully for the same purpose, and we later used it to extend our measurements on Byrd Glacier. However, aerial photographs can only be used in photogrammetry if their scale is known, and in the Antarctic at that time we started without maps or ground control from which to determine scale. Ground survey therefore

made our stay at each glacier worthwhile, quite apart from the many other observations we were able to make. The whole area was untrodden before our little party reached it.

The helos came again on 29 November to move us 80 miles down the coast to Barne Inlet. Tuck came, and for the first time we had our full complement. Krebs had brought no chart and asked, "Do you know where this place is?" It was a fine day and we could see 200 miles north and south, so there was no problem. We landed at the foot of Horney Bluff, a small pyramidal peak on the left bank, overlooking the ice 3,700 feet below. Behind us rose the 11,000-foot ice-clad summits of the Britannia Range. In front lay the immense glacier. We were overwhelmed by the spectacle, knowing that neither Scott nor Shackleton were even aware of its existence. To us it was still Barne Inlet, even on the latest maps. The New Zealanders eventually named the glacier for Richard Evelyn Byrd.

After planting seven stakes in a line across the glacier, we faced a major disappointment: Taylor had not seen a single stake in the theodolite, and had even lost the helicopter from view. Such were the distances involved that, even with his 30-power telescope, the helicopter had melted into the chaotic background of séracs and crevasses.

With their work done, the helicopters flew back to McMurdo Sound that evening. We spent three days and three nights searching for the stakes in the telescope. We measured a 6,000-foot baseline and were ready for one final effort when the helicopters reappeared on 2 December.

Picnic lunch beside Byrd Glacier. I was lowered onto
the ice from the winch over the door of the helicopter.

First came lunch. The aircrew had brought a whole cooked ham. We carved it on the lid of a fuel drum and, missing only some deck chairs, enjoyed an incongruous picnic in brilliant sunshine. The visitors brought with them an extra theodolite and two portable radios. With a theodolite and radio at the two ends of the baseline, we were able to see and to speak to the pilots as they flew over the glacier to search for the stakes. They found three, hovered over each in turn and threw smoke bombs while Taylor and I, at opposite ends of the baseline, read angles to fix their position. It was an eminently successful operation, though we could have hoped to find more stakes. The helicopters climbed back and rapidly loaded our camp and survey gear. Jack and I flew with Hickey along a hidden valley to enjoy the spectacle of a giant icefall tumbling fully 10,000 feet down the face of Mount Aldrich. Two hours later we were at McMurdo, 200 miles away.

The next trip, on 7 December, was to Cape Crozier at the eastern end of Ross Island, to find and survey the stakes that Sven Evteev and I had set up in February. I flew with Lieutenant Steven (Steve) Snyder, USN, stowing Taylor and Darby in the cabin below. John Hickey flew a second helicopter carrying Brian Roberts, Hugo Neuburg, and Donald (Curly) Wohlschlag. Roberts, who was in the Antarctic that season as official British observer with Operation Deep Freeze, was an old friend. He was a tubby, jovial man who had been on an Antarctic expedition in the nineteen thirties and now occupied the "Antarctic desk" at the Foreign Office in London. A zoologist by training and an ornithologist by inclination, he had an encyclopedic knowledge of all things Antarctic and had played a pivotal role in setting up the 1959 Antarctic Treaty. As I had done earlier on many a flight, these three were hitch-hiking, this time in the hope of visiting the emperor penguin rookery at the foot of the cliff below our camp site. Being equally curious, we too took the opportunity to see the penguins. So all piled into Hickey's machine and dropped to the sea-ice below. There were more than a hundred grey, molting emperor chicks, chaperoned by a few adults. Some adults were striding purposefully towards the open water in the distance but evidently most were out at sea fishing. The parents must have been diligent, because nearly all the chicks were fat, a few showing small patches of white breast where their grey downy feathers were beginning

Little Horrible unloading at Beardmore weather station.

to fall away.

With a penchant for pleasing journalists — forty-four of them came to McMurdo during the season — the management had sent along a photographer. While we had landed at a distance and walked quietly towards the emperors, now the helicopter approached closer for the benefit of the photographer. It hovered near enough and long enough to drive all the chicks into panicked retreat, a bunch of them huddling at the top of a snowdrift leading up to the ice shelf. My diary notes: "God knows what effect this will have on their breeding habits."

After depositing our party at the cliff top the helicopters returned to McMurdo. We spent two days re-fixing the position of the stakes on the ice shelf, finding to our amazement that the one nearest to the land had moved at the rate of 1,600 feet-per-year, the furthest 2,264 feet.

Later, at a press conference at McMurdo, I had reason to need these figures to respond to an unexpected question. Some of the scientific staff at McMurdo were required to present themselves before a couple of dozen assembled journalists to face a barrage of questions. The press people were supposed to be taking an interest in science, but when they considered our interests esoteric, their minds wandered towards their own interests. One ghoulish reporter asked, "Where are the bodies of Captain Scott, Edward Wilson, and Birdie Bowers?" He could see little use for glaciology except to answer practical questions, and he was persistent. I knew that Shackleton had determined the rate of movement off Minna Bluff not far from the polar party's last camp. Interpolating between Shackleton's measurements and my own, and counting the 52 years that had elapsed, I jabbed a finger at a point on the wall map, adding that the bodies would now be under some 100 feet of snow. When he continued with "Could we dig him up?" I said it was time to move on to the next question.

The press conference ended with most questions answered and I steered my mind back to the next task. We needed another set of angle measurements at Byrd Glacier before the stakes were lost forever. This we did on 9 December, confident that a stream of its dimensions would move a measurable distance in seven days. John Hickey flew at 9,000 feet over a pass between Mount Morning, an extinct volcano, and Mount Cocks, an 8,000-foot peak in the Royal Society Range. Brian Roberts came along to record angles for me. The pilots called it a CAVU (Ceiling and Visibility Unlimited) day, the sort you remember long after all the cloudy and windy days have been forgotten. Peering through our instruments, we were relieved to find all three stakes. It is characteristic of the age in which we live that, in a single afternoon, we flew to a mountain peak 200 miles from McMurdo, made a vital set of observations, and returned to base for dinner. Perhaps more astonishing were the results: Byrd Glacier had moved at the rate of seven feet per day (half a mile per year) since our last visit. Unbeknown to us at the time, this was to prove the fastest-moving glacier flowing through the 2,000-mile chain of the Transantarctic Mountains.

---

[1] Paul Siple. *90° South. The Story of the American South Pole Conquest.* New York, G. P. Putnam's Sons, 1959.

2 Edward Shackleton. *Nansen the Explorer*. London, Witherby, 1959.

3 Alan Reece. Sledges of the Norwegian-British-Swedish Antarctic Expedition, 1949–52. *Polar Record,* Vol. 6, No. 46, 1953, pp. 775–787.

4 Charles W. M. Swithinbank, David G. Darby, and Donald E. Wohlschlag. Faunal remains on an Antarctic ice shelf. *Science,* Vol. 133, No. 3455, 1961, pp. 764–766.

5 J. D. Back (ed.). *The Quiet Land, the Diaries of Frank Debenham.* Huntingdon, Bluntisham Books; and Harleston, Erskine Press, 1992.

6 F. Debenham. A new mode of transportation by ice. *Quarterly Journal of the Geological Society,* London, Vol. 75, Part 2, 1920, pp. 51–76.

7 I. A. Zotikov and A. J. Gow. The thermal and compositional structure of the Koettlitz Ice Tongue, McMurdo Sound, Antarctica. In: International Conference on Low temperature Science, Sapporo, Japan, April 14-19, 1966, *Physics of Snow and Ice: Proceedings,* Vol. 1, Part 1. Hokkaido, Japan, Institute of Low Temperature Science, 1967, pp. 469–478.

8 Charles R. Wilson and A. P. Crary. Ice movement studies on the Skelton Glacier. *Journal of Glaciology,* Vol. 3, No. 29, 1961, pp. 873-878.

9 *New York Herald Tribune,* 12, 14, and 16 December 1960.

Chapter Five
Mechanical Dogs

Yea, though I walk through the valley of the shadow of death, I will fear no evil.

Psalm 23

The rest of the big glaciers were said to be beyond helicopter range from McMurdo. So the following day we drove our two Eliason motor sledges down to the sea-ice runway with about two ton of camping equipment, sledges, food for three months, aluminum poles, and instruments. The frail-looking vehicles gave rise to some merriment among the Navy people, whose natural affinity was for giant tractors weighing many tons. Half a century had elapsed since anyone had tried motor sledges in the Antarctic, and there were grim prognoses about their performance. Waiting for us was an R4D, really just a good old trusty twin-engined DC3 in Navy colors and on skis. On her nose was painted the name *Little Horrible*. The aircrew scowled at the size of our load — I think aircrews always scowl at loads. However, they put it on board. It reached the ceiling except where a space remained on top of the cargo for the four of us to lie prone. I had allowed Tuck to fly to Byrd Station to satisfy his interest in the survival of buildings erected in snow trenches.

So in his place we had Olav Liestøl, a Norwegian glaciologist visiting the Antarctic for some weeks as official observer for his country. Born with his skis on, like most of his countrymen, I knew that he would not let us down.

The crew used eight JATO rockets to get the heavily loaded plane off the ground. Having leapt into the air as usual, this time the plane began to lose altitude, and I could see that the pilots in "the front office," as they call the cockpit, were concerned. It took nearly half an hour to gain enough altitude to clear the 2,300-foot saddle of Minna Bluff. Shortly after this, all four of us were summoned to huddle up front because the fuel used had made the machine tail-heavy. Civil aviation authorities would have been horrified, but in this environment I trusted the pilots to balance the risks of overloading against other problems which could arise if we halved the load but doubled the flying hours.

After three hours we landed at the remote Beardmore weather station. As on

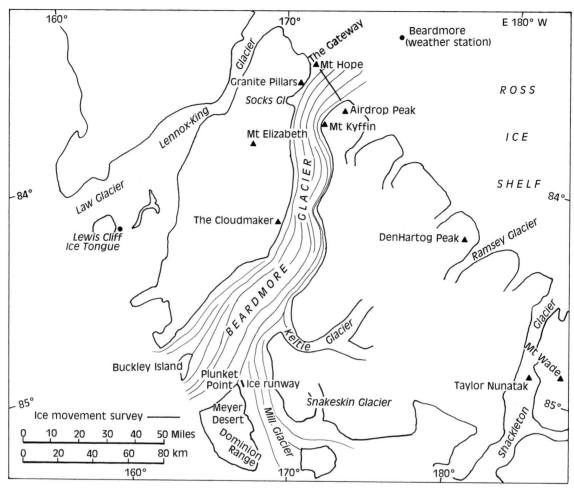

Beardmore Glacier.

our last visit, the station consisted of three men in a Jamesway hut. Later, we came to refer to the place as The City, because it was the largest settlement on the Ross Ice Shelf and served as our advanced base for several months. It was a calm, warm day with the thermometer reading +25° F, warmer probably than our families were encountering in the mid-latitudes of the Northern Hemisphere. Before us lay a breathtaking panorama, 200 miles of high mountains stretching from Mount Wade in the south to Cape Goldie in the north. We unloaded and set up camp, sorting the loads for an early start towards Beardmore Glacier the next day.

Considering motor sledges to be the mechanical equivalent of dog teams, we had brought standard dog-sledging equipment and supplies: Nansen sledges, pyramid tents and conventional pemmican-based lightweight rations. But it did not take long to discover that for some Americans, the word ration conjures up images of privation, gnawing hunger, and the heroic age of exploration. Heroism was the last thing we needed, so I explained that what I meant by a sledging ration is a planned diet that — to save weight — avoids foods containing water. There was plenty of water, albeit in solid form, right under our feet. Ordinary canned foods should be avoided because they contain more water than solids. I had no intention of limiting our intake of food. Our range of travel depended critically on how much we carried. But we were to be working outside, within a few hundred miles of the South Pole, and facing all kinds of weather. A nourishing and satisfying diet was essential.

We had packed the food into narrow boxes each containing everything necessary for twenty man-days. That meant that one box in a two-man tent should last about 10 days. Some would find that it lasted nine days, others 11 days. The convenience of having everything on hand is far preferable to collecting the ingredients of each meal from boxes outside. Anyone coming into a tent brings with him a certain amount of snow clinging to clothes and boots. The snow melts and drips onto sleeping bags, so the fewer excursions the better. The need to balance protein, fat, and carbohydrate left little room for maneuver. Our ration was a direct descendant of rations used on polar expeditions since the turn of the century, though it was adapted to the availability of new foods and had to use standard package sizes. This necessitated minor departures from what might be considered ideal.

The only item that one would not find in a good food store was pemmican. North American Indians have made pemmican for centuries by drying buffalo or deer meat and pounding it into a powder. The powdered meat is then mixed with hot fat. When the mass of fat and meat cools, it is cut into cakes. Our pemmican was made from dehydrated minced beef (55 percent) and pork (45 percent) compressed into three-ounce blocks, and to avoid any imagined association with privation, it was labelled "Meat Bar." The daily ration — or planned diet — had an estimated food value of 4,700 calories. It came out like this:

Item	Ounces per man-day
Meat bar	4.5
Butter	2.0
Wholemeal cookies	2.0
Rolled oats	2.0
Raisins	1.2
Whole milk powder	2.0
Granulated sugar	2.4
Hot chocolate mix	3.2
Instant mashed potatoes	1.0
Dehydrated onions	0.8
Instant coffee	0.4
Rich fruit cake	3.2
Marmite	0.4
Seasonings	0.2
Fruit juice crystals	1.6
Strongly flavoured cheese	0.8
Cadbury's fruit and nut milk chocolate	4.0
Total weight per man-day (ounces)	31.7

Breakfast consisted first of fruit juice, then oatmeal[1] liberally laced with butter, sugar, raisins, and milk powder, followed by coffee. Lunch was cookies, a bar of chocolate, and a vacuum flask of fruit juice, coffee, or drinking chocolate. Although I prefer plain (dark) chocolate, Antarcticans learn not to try it. Rock hard in the cold, it can break teeth. Supper was pemmican stew laced with butter and onions and thickened with potatoes, and more fluids. The fruit cake was pure indulgence — a delicious luxury. The fruit juice crystals are a great invention, made of real dehydrated fruit juice rather than just citric acid flavoring. Water derived from Antarctic snow is purer than distilled water back home, hence rather tasteless. Dehydration is a common problem for people working in cold climates. We needed to drink several quarts every day to compensate for the moisture lost through breathing. The fruit juice made drinking a pleasure instead of a duty.

We had begged four pairs of shiny new Head skis from the manufacturers, who generously obliged. Although we did not expect any downhill skiing, I have always favored downhill skis because they are wider than cross-country skis. This spreads the load, making them safer for crossing weak snow bridges. In addition to all necessary camp and survey gear, we had to carry 15 large aluminum pipes to mark ice movement stations on each glacier. So we found it necessary to haul two 13-foot Nansen sledges behind each motor sledge. Though the Eliason can seat two men, the weight is concentrated on a fairly small area and I was worried about driving over crevasses. We had to spread the load to reduce the risk of collapsing weak snow bridges. Aerial photographs showed that the mouth of the Beardmore was badly crevassed, and the descriptions of it by Shackleton and Scott were fresh in my mind. We experimented with remote control of the Eliasons by means of long ropes, for though I was prepared to risk a vehicle, there was no need to risk lives. I told a Navy friend that my machine could do everything that a dog could do except wag its tail.

When our little party, after two days of poor weather, finally set course for Mount Kyffin at the mouth of the glacier, our caravan comprised two sledge trains, each spread over a distance of 60 feet. In the lead were the two riderless motor sledges, intruding upon the Antarctic silence with the unfamiliar putt-putt sound of lawn mowers. Behind each, at the respectful distance of 12 feet, followed a loaded sledge, and 12 feet behind that, another. A sledge wheel, consisting of a bicycle wheel with a mile counter, was attached to the rear sledge to record the distance traveled. Finally came two men on skis, one steering the train and being pulled by two 60-foot nylon ropes attached to the steering skis of the Eliason, the other being towed by a rope trailing

We drove by remote control from 60 feet behind to reduce the risk of falling into crevasses.

behind the last sledge. None of us sat on the machines except to start them. Since that day we have driven thousands of miles without altering the system of control, having found that riding on skis is not only much safer but also more comfortable than the unsprung chassis of the vehicle. It is, moreover, less noisy, and thus more restful. Our caravan glided over the snow at four knots, giving us plenty of time to admire the scenery.

Beardmore Glacier is 16 miles wide at its mouth, and we proposed to set up a line of stakes between Mount Hope and Airdrop Peak. There was only one possible choice of line between crevasse fields. Opposite Mount Kyffin the glacier is narrower, but there is a six-mile-wide icefall where the stream tumbles round a bend in the valley. We came into crevasses when still about 25 miles from Mount Kyffin, and as the light was bad, we made camp. The sun never sets in summer, so bad light means overcast conditions and a lack of shadows. Surrounded by crevasses, invisible because of whiteout conditions, we felt obliged to pitch a tent without unfastening our skis. Inside the tent we stretched out our sleeping bags on up-turned skis. It was an inauspicious start to the expedition.

The next morning, 14 December, I awoke at 0200 to find the sky clear, and an hour later we were off. After a mile we came into a two-mile-wide relatively crevasse-free channel that we had seen in the aerial photographs, the only possible approach route from Beardmore Station. This led into a valley just below the icefall; here we turned north to head for Mount Hope. The last two miles were unpleasant. We found that every one of these big glaciers, even though relatively crevasse-free in the middle, had a one- to two-mile-wide crevassed zone in a shallow depression at each margin, evidently due to the rapid shearing that takes place there.

We probed some snow bridges, twisting and turning to avoid the most treacherous areas. The performance of the motor sledges was amazing. Many times the track, which is the heaviest part of the vehicle, left a long black hole in a snow bridge. But each time, one of the track lugs caught on the far lip of the crevasse and the Eliason climbed out by itself. The advantage of being 60 feet behind the machine through this maze of crevasses was that we were always ready to let go the ropes if the machine should drop. On the other hand, once the vehicle was across a given hazard and we ourselves were approaching it, we hung on for dear life with the intention of being pulled out if our skis broke through. At the end of long ropes we were, moreover, free to ski clear of the black holes left by the motor sledges, instead of being dragged inexorably across them as were the Nansen sledges. It was with relief that we reached safe ground and approached the Gap, as Shackleton called the low col which separates Mount Hope from the mainland. At this point *Little Horrible* appeared overhead to see that all was well, dipping her wings when the crew caught sight of us. We reeled out the antenna of a small battery-powered radio and exchanged greetings with them. A solitary skua circled overhead and we marvelled at its ability to fly four hundred miles inland without any possibility of finding food along the way.

The mouth of Beardmore Glacier from an altitude of 10,000 feet. Shackleton discovered the glacier on 4 December 1908 from the summit of Mount Hope (center). He approached from north (right), passing through the gap (now The Gateway) between Mount Hope and the mainland. Our route was just off the bottom of this photograph.

Shackleton discovered the Gap, later known as The Gateway, on 4 December 1908 and the same day climbed Mount Hope. From the summit he saw a giant glacier, which he named Beardmore for one of his financial sponsors, trending almost due south. It was the key to his reaching the polar plateau and his farthest south, less than 100 miles from the South Pole, on 9 January 1909.[2] The Gap was later the site of a cache laid by the Ross Sea Shore Party of Shackleton's ill-fated

Imperial Trans-Antarctic Expedition of 1914–1917.[3] Not knowing that Shackleton's ship *Endurance* had been sunk in the Weddell Sea, which precluded any attempt at crossing the continent, the party established the food cache, as planned, in the Gap on 26 January 1916.[4] One member of the party died of scurvy during the return journey to McMurdo Sound. We were the first people to visit the Gap since the cache was laid. From the low col we searched long and hard with binoculars but had to conclude that everything had been buried by snow in the intervening 44 years. In 1963 I corresponded with R. W. (Dick) Richards, the only survivor of the party, saying I was sorry not to have found some token of his party's very gallant achievement.

One of the Eliasons broke down and it took us some time to find the trouble. Of the four of us, Taylor was the best mechanic, and he discovered a broken valve spring. This is a rare ailment in a good engine, and though we carried a multitude of spare parts, we had brought no valve springs. There was plenty of time to contemplate our predicament, for the next five days brought only fresh snow, blowing snow, and overcast conditions in which we could not safely proceed. My diary records reading Arthur Koestler's *Darkness at Noon*, John Steinbeck's *Of Mice and Men*, and a book about Marco Polo. Although we had a daily radio schedule with Beardmore Station, it was a comfort to find, when we called about the valve spring, that the operator 400 miles away at McMurdo was the first to acknowledge my faint calls on the Morse key.

Finally on 20 December the sun shone. We put everything necessary on three sledges behind the surviving Eliason and headed for the southeast corner of Mount Hope. Though there was a wide and deep windscoop protecting the place like a moat, we got across and made camp among some boulders of a lateral moraine. Liestøl climbed 2,000 feet up the mountain to find suitable survey stations. The surface of an active glacier is generally slightly convex; so he had to climb high to see the far side of the glacier. We set off to lay stakes as soon as Liestøl returned, leaving Taylor alone to climb with the theodolite and a small radio. Crossing the marginal crevasses, we learned that it is disconcerting to peer down the black holes which punctuate the trail of a motor sledge, because the bottom is invisible.

For a short distance in the worst area, I reconnoitred ahead on skis. There was a moment that I shall never forget. A snow bridge collapsed under the middle of my skis. The ends of the skis — front and back — were on safe ground, but I myself was poised over a menacing black hole. Momentarily unbalanced by the sudden sag in the middle of the skis, I steadied myself with ski poles. Now that I was bridging the crevasse, the problem was what to do next. Lifting one ski would put all my weight on the other, which might break. Very carefully, I put weight on one ski pole placed on the far lip, and more weight on the other placed behind. Transferring the least possible weight from one ski to the other, I slid it forward onto the far side. Then I put all my weight on that ski and brought up the other. I was sweating, and in shock — but I was safe.

In the middle of the glacier we were running over windslab, a hard snow crust that breaks under the weight of sledges and skiers. We could see that there were crevasses about, so every time the crust gave way we had no way of knowing whether or not it was the start of a free fall. As it turned out, the drop was probably a couple of inches, but it added some white hairs to our heads. We covered 34 miles and set up 12 stakes across the glacier, contacting Taylor at every hour on the radio to make sure that he had seen them. Back in camp we were almost too tired to eat. Taylor had been standing by the theodolite in a rising wind for 11 hours.

Now I realized what an exceptionally keen field team we had assembled. My colleagues were never the first to suggest calling it a day — I was.

We spent the next five days measuring a baseline and doing all the other observations that came to be a part of our routine at each glacier. I took angles to each stake while the others climbed Mount Hope, the first party to stand on the summit since Shackleton and his companions in 1908. The object of making more than a single set of observations was to establish a first approximation of the rate of ice movement. By extrapolation we should know where to look for the stakes the following year. Without this information, trying to find a four-inch-diameter pole at anything up to 14 miles after a year's ice movement would border on the impossible. The sun shone on Christmas day, so we struck camp and headed back across the crevasses. Liestøl went ahead on skis; we followed with our one poor motor sledge hauling three loaded Nansen sledges and three men.

Darby was not yet confident about maneuvering his skis between the crevasses. In the circumstances, we felt that the safest place for him was to sit like a camel driver astride the load of a Nansen sledge. With the total load approaching one ton, the Eliason answered only slowly to the helm, and at one stage the whole train was cruising along a snow bridge with nothing even touching safe ground. However, the crevasses were kind that day, and we were able to stop in the middle of the glacier. Even so, we at first inadvertently chose the bridge of a colossal crevasse for a camp site, and hurriedly moved on when someone noticed the sag in the snow surface. We pitched the tent only after roping up and doing a sort of war dance in ski boots to test the integrity of the snow. The experience paralleled one that Shackleton had not far from here. His diary for 7 December 1908 reads:

> *When we tried to camp to-night we stuck our ice-axes into the snow to see whether there were any more hidden crevasses, and everywhere the axes went through. It would have been folly to have pitched our camp in that place, as we might easily have dropped through during the night.*[5]

The only things we had to celebrate Christmas were a few party balloons, brownies from home, some extra candy and a tot of whisky. The 30 miles to Beardmore Station were covered the following day.

On finding that the weather station radio was not in working order, we invited the Navy men into our tent to use our radio to call for spare parts. One of the rallying cries of the Navy public relations staff at McMurdo was SEA POWER SUPPORTS SCIENCE. It gave us some satisfaction to remind them that here in the backwoods, SCIENCE SUPPORTS SEA POWER. They laughed as much as we did. George Toney had sent with the last plane a bottle of gin, a bottle of whisky, and a bottle of cognac, so science and sea power collaborated late into the night.

Our next glacier was the Nimrod, 120 miles to the northwest. Though it was stretching their capabilities, we had asked for helicopters to carry us from Beardmore Station. However, we now learned that there was no immediate prospect of any aircraft arriving from McMurdo, so we prepared to start with the one motor sledge hauling all our equipment, provisions for six weeks and 32 gallons of motor fuel. Tuck was stranded at McMurdo, and Liestøl decided to wait with the Navy men for a flight to McMurdo and home. We were sorry to lose him but pleased to have tempted him away from his proper job of observing the whole USARP operation. Taylor, Darby, and I got under way on 29 December, traveling on the ice shelf parallel with the trend of the mountain range. Each time we stopped to fill the fuel tank, we drilled in an aluminum marker and resected its position by taking angles to mountain peaks. These extra measurements would show how fast the ice shelf was moving.

Including our own weight, the little machine was hauling 2,300 pounds. For a vehicle designed to carry just two people, there are few kinds of abuse more cruel than overloading. We expected a breakdown at any minute. However, six days passed, and on the seventh we were 20 miles off Shackleton Inlet at the mouth of the Nimrod Glacier. We approached with caution. In 1902 a sledge party comprising Scott, Shackleton, and Wilson had been prevented from reaching land in this latitude by " . . . an enormous chasm filled with a chaotic confusion of ice-blocks."[6] The light was bad and we did not want to find the chasm by falling into it. By heading at right angles to the mountain range instead of cutting the corner from the south, we were able to pass between two great marginal strings of crevasses and to climb the middle of the glacier without knowingly crossing any unpleasant areas.

We made camp at a point six miles from Cape Lyttelton, and at once began to reconnoiter a ski route towards the right bank of the glacier where there appeared to be a climbable hill. Just as we had found on the Beardmore, here also there was a mile-wide depression riddled with crevasses. When we reluctantly became accustomed to picking our tracks across this sort of terrain, we called it the Valley of the Shadow. Here, it seemed, life and death were bedfellows. We were in a state of mild shock after each crossing.

We prayed for helicopters, but the radio told us that they were busy elsewhere. The next day we climbed the hill, roping up to traverse an icefall that barred the way, and found ourselves 2,000 feet above the glacier. After marking the ends of a good baseline we returned to camp, drilling in three movement stakes on the way. There followed four days of poor weather, during one of which our camp was struck by a curious series of hurricane-force winds interspersed with calms lasting for a minute or so. It was apparently an eddy effect caused by the swirling of a southerly storm over the mountains to the south. We feared for our tents, having oriented them in the expectation of a down-glacier wind.

On 10 January we planted the remaining stakes in a line across the glacier, and the following day fixed their positions from the hill stations. The weather was clear, and we were amazed to hear the sound of running water on the hillside. Though it was under the snow, we dug to find a babbling stream of meltwater. The explanation seemed to be that the wind kept the snow surface cold, whereas sunlight penetrating to the rock beneath warmed it to the melting point. We drank heartily.

After completing sundry observations, we reached camp at midnight, 16 hours

Nimrod Glacier (looking upstream) from an altitude of 19,000 feet. Kon-Tiki Nunatak divides the glacier and then makes its mark on the confluence. The glacier is eight miles wide at its narrowest point. We crossed in the foreground.

after starting from it in the morning. On these long absences from camp the weather was a constant anxiety, because the mere arrival of a continuous cloud cover could make both our tracks and the crevasses virtually indistinguishable from the rest of the surface.

Having been led to expect that helicopters would pick us up by the time we reached Nimrod, we now faced a fuel shortage for the unplanned return journey to Beardmore Station. The Eliason had used a little more than half our supply to cover the 159-mile outward leg. Allowing for the decreasing food and fuel load, and with 10 great aluminum pipes left behind on the glacier, we still had no fuel reserve. We discussed the unwelcome prospect of man-hauling, but determined to postpone the event until our last drop of fuel was spent. Finally, after five days of traveling and three days of lying up because of bad weather, we caught sight of the hut at Beardmore Station in a shimmering mirage when still 18 miles out. I have learned that mirages are as common in cold deserts like the Antarctic as they are in hot deserts. Four hours later the gallant little machine hauled us home in the face of a rising gale. When we stopped the engine just after midnight on 19 January, the visibility was only 100 yards. There were a few pints of fuel left in the tank.

Tuck was there to greet us, having waited 17 days for our arrival. We were by this time thoroughly impressed by the performance of the motor sledge. It had hauled an average of 2,000 pounds over a distance of 320 miles at the expense of 36 gallons of fuel. This meant that the potential unsupported range of the machine, hauling a load consisting mainly of its own fuel, was in excess of 2,000 miles. The possibilities were immense, and we began at once to plan a trip to the Shackleton Glacier down the mountain range to the southeast. But radio messages daily reported the imminent arrival of two helicopters, and the possibility of saving another three-week sledge journey was too tempting.

We spent three weeks in local survey work, finding from the ice temperature at a depth of 33 feet that the mean annual air temperature hereabouts is -14° F. The rate of snow accumulation is conventionally reported not as the depth of snow added to the surface but as its water equivalent if you melted it. The density of snow varies, so that just measuring its depth would not allow a true comparison between different places. We found that the surface at Beardmore Station was being added to at the rate of 7.7 inches of water per annum — little more than in parts of the Sahara Desert. Throughout December and January, daytime air temperatures remained between +18° F and +28° F, much warmer than we expected to find here, within 500 miles of the South Pole.

We had "baths" with a basin of water each on the tent floor, our first ablutions for seven weeks. I generally tried to wash and put on clean clothes once a month. Surprisingly, even after weeks in the same clothes, we rarely noticed our tent companion's body odor. Perhaps two bad smells cancel each other.

One day when we were confined to the tents by weather, and Darby and Tuck were spending their time in the Jamesway hut with the three sailors, I heard a shout.

"Charles, come quick, there's something wrong. Stop the generator!"

The author (left) with David Darby, Jack Tuck, and Tom Taylor.

He was referring to the diesel generator that powered the lights for the hut. Taylor and I quickly got out to find Darby staggering towards us, supported by Gary Signor, one of the Navy men. We had an inkling of what might be wrong. Running through the hut we saw Jack Tuck lying on a camp bed with a bemused expression on his face. The downwind door was blocked by snow. We ordered Tom Badger, the second Navy man, to stop the generator, which he did, though we noticed that he was staggering and seemed confused. We then opened all the doors and windows.

At this point the four occupants reported having severe headache and nausea. Badger was stooped like a hunchback and his hands were shaking violently. Feeling on the verge of fainting, he had left the hut shortly before. Now exposed to fresh air, all of them reported having had headaches and a loss of appetite at supper, though none had admitted it. Two of them had quietly taken aspirin. After supper Tuck lay down while Badger and Signor, each feeling the need for air, staggered outside. At this point none of them suspected carbon monoxide poisoning from the diesel fumes. Only when Darby followed them, and Signor remarked on his staggering gait, did Signor admit that he too was feeling ill. Darby then understood that the situation was serious.

My diary entry for the day includes this paragraph:

> *I came quite close to having two dead USARPS and two dead Navy men on my hands. It is an old, old story. An internal combustion engine should never be in or attached to a living hut. There should always be a ventilated space between. Since the same mistake has been made so often in the past, the man who designed or permitted to be built a place like this ought to have known better. The occasion, too, was typical. Bad weather, windows shut to keep out blowing snow, generator kept running for long hours just for lighting and to keep music blaring from the radio.*

The best-known story of carbon monoxide poisoning in an Antarctic hut is told by Richard Byrd, who lived alone under snow on the Ross Ice Shelf.[7] I once experienced it in a tent in Iceland. The insidious feature is that, with dulled senses, one does not understand what is happening.

The helicopters arrived on 28 January but the summer weather was coming to an end and we managed only to pick up the disabled Eliason at Mount Hope. The job did not go smoothly. The motor sledge projected from the side door, preventing it from closing. The Nansen sledge was hung by its traces beneath the helicopter. Shortly after taking off there were two unaccountable bangs followed by a thud which shook the whole machine. I guessed that, blown by the slipstream, the sledge was hitting the tail. Acutely aware of the consequences if the sledge hit the tail rotor, I turned to Hickey and yelled "Drop it!" To my great relief he did, and we landed to mark where to find it later. The 100-foot drop had not done irreparable damage.

Back at the weather station, an R4D flew over just as fog was rolling in. It turned to find that its own contrail — the condensation trail from the engines — had obscured the landing area. Landing on instruments, they taxied some miles in fog while searching for us on radar. It took a search party on Eliasons to find them.

With long delays due to poor weather, the helicopters were ordered back to McMurdo. Some of the pilots had an unhealthy preference for their beds at McMurdo as opposed to sleeping in a tent. Steve Snyder was one of them. Although we were in virtual whiteout with low cloud, he took off to fly home with his engineer and one passenger. After 100 yards the helicopter flew into the ground, its wheels digging a long trench as it came to a halt, denting the engine cover. Snyder was unable to explain what had happened, but it was clear enough to those watching. There was ice on the rotor blades and no visible horizon. After recovering from shock — with coffee — he took off again. We heard later over the radio that he had to give up and camp half way to McMurdo. My diary entry was coldly unsympathetic:

> *Serve them right for trying to stretch their luck in marginal weather. They will spend tonight shivering in the helicopter and cursing everybody and everything (for their own blunders).*

But we too made mistakes — and

learned by them. Some were simply embarrassing — good for a laugh. Others brought awareness that the environment could be unforgiving.

We were picked up on 17 February by a Hercules. Our baggage weighed around three ton but nobody complained. Later that day, propping up the bar in the Officer's Mess at McMurdo, Krebs reported that during the season we had used a total of 91 hours of helicopter flight time. I could not help wondering what Shackleton would have made of it.

Four days later we were in Christchurch, New Zealand. In the Gainsborough Hotel I peeled off my Antarctic clothes and put them in the garbage can.

---

[1] Oatmeal is an ideal breakfast food. It is a polysaccharide and hence the carbohydrate within it is slowly absorbed, an effect which is further increased by the presence of a gum found in oats. The slow absorption provides energy throughout the morning and discourages the consumption of snacks.

[2] E. H. Shackleton. *The Heart of the Antarctic, being the Story of the British Antarctic Expedition 1907–1909.* London, William Heinemann, 1909 (2 vols).

[3] E. H. Shackleton. *South. The Story of Shackleton's Last Expedition 1914–1917.* London, William Heinemann, 1919.

[4] R. W. Richards. The Ross Sea Shore Party 1914–17. *Scott Polar Research Institute Special Publication,* No. 4. Cambridge, Scott Polar Research Institute, 1962.

[5] E. H. Shackleton. *The Heart of the Antarctic, being the Story of the British Antarctic Expedition 1907–1909.* London, William Heinemann, 1909 (Vol. 1, p. 316).

[6] R. F. Scott. *The Voyage of the Discovery.* London, Smith Elder & Co., 1905 (Vol. 2, p. 47).

[7] Richard E. Byrd. *Alone.* London, Putnam, 1938.

Chapter Six

Fatal Accident

Lector, si monumentum requiris, circumspice.
Christopher Wren (1675–1747)

I spent the northern summer of 1961 at the University of Michigan organizing the next expedition. At McMurdo we had left all the basic equipment needed to finish the glacier movement project, so now I could spend time on improvements. The next field party consisted of Arthur Rundle, a young geographer from the University of Durham in England, and Tom Taylor. Tom and I had squabbled on a few occasions the previous season, so it was a delightful surprise when USGS told me that he had made a specific request to be assigned to my party once again. I had come to have a great respect for his common sense and good judgment, quite apart from his stamina and willingness to work under any conditions. He was not one to complain of hardship, because in his eyes, nothing that we did could be called hardship. Though neither of us would ever make a diplomat, Tom had an insight into the mind of government servants that was based on long experience. He was a staunch ally throughout two long field seasons.

Rundle was commended to me by Hal Lister, glaciologist with Sir Vivian Fuchs' Trans-Antarctic Expedition of 1957–1958 and Lecturer in Geography at the University of Durham in England. A trim and cheerful Yorkshireman, slow to chide and swift to bless, Arthur was horrified to discover that there were no pubs in Ann Arbor, only some rather sleazy bars. How could an Englishman survive in such an uncouth world? However, as often happens with immigrants, the initial shock was soon replaced by enthusiasm for the relatively high standard of living — to which he rapidly became accustomed. He confided a lifelong ambition to own a Buick and drive it home to Durham. But he never did. Instead he later settled in Columbus, Ohio, and with a Ph.D. in glaciology, lived happily ever after.

Our sledge journey to the Nimrod Glacier had shown that three men could do the work and we had grown confident that three offered a sufficient safety factor. The first task was to revisit the glaciers studied the previous year to find how far the stakes had moved. However, in order to calculate the volume of ice discharged into the ice shelf, we now needed to know the ice thickness. In fact we wanted a complete cross section at each stake line. Seismic soundings and gravity methods were the best available means of obtaining this. I therefore approached Edward C. (Ed) Thiel of the University of Minnesota, one of the most experienced of Antarctic geophysicists, and proposed that we should work together. He agreed, and brought as assistants James (Jim) Olson and Harold (Hal) Linder, both Ph.D. students from his university. He also brought a twelve-channel exploration seismograph and a Worden gravity meter. Tom and I agreed to introduce him to the art of motor sledging, and for the purpose acquired a new vehicle, this time of US manufacture. It was a Polaris, built in Roseau, Minnesota, similar to the Eliason but rather more sophisticated in design and heavier.

We had also found time to refine some of our camping and trail gear. To lighten the fuel load on the sledges we introduced plastic jerrycans of a type guaranteed not to contaminate gasoline nor to dissolve in it. As an experiment we also took a rubber 55-gallon Sealdrum for bulk gasoline. Shaped like a steel fuel drum but soft-sided, it would ride on the sledge with less chance of breaking anything. Made-to-measure aluminum boxes held everything in place on the sledges. Copper-bottomed stainless steel saucepans replaced the traditional aluminum pans, and hard plastic mugs and bowls replaced enamel plates and cups. Though visitors to the tent never knew it, the result was that we went through the whole season without ever washing dishes. The high fat content of our diet and the absence of burnt food sticking to the pans meant that we could clean everything by wiping it with toilet paper. Water on the trail is expensive; every drop has to be melted from cold snow. Wiping saved fuel. However, there is a residual ethic in the sledging fraternity that favors primitive

Arthur Rundle.

and inconvenient cooking utensils. Thus our eminently successful improvements never caught on outside our little group. Burned aluminum, cracked enamel mugs, crumbling pot scourers and dirty dish water are evidently the order of the day for macho explorers.

The "cook" box for each two-man tent was made of aluminum and measured 10 x 10 x 24 inches. When full, it weighed 25 pounds. It belonged in the middle of the tent, parallel with and between the sleeping bags. In it we kept:

> One Coleman Sportsmaster single burner stove
> One four-quart stainless steel saucepan with lid
> One three-quart stainless steel saucepan with lid
> One two-quart stainless steel saucepan with lid
> Two tablespoons
> Two one-pint hard plastic mugs
> Two large hard plastic soup plates
> One knife
> One can opener
> One box matches
> One Coleman filter funnel
> One roll toilet paper
> One snow brush (for clearing the tent)

Tablespoons were the only utensils needed for eating or stirring.

We took eider-down suits for the whole party. Standing still on mountain peaks for hours on end had shown that the standard issue parka left much to be desired. Gloves presented another problem. Handling a theodolite requires fine tuning of small knobs with delicate fingers. Heavy mittens keep you warm but are too clumsy. We had to use leather finger gloves up to the point that the cold killed all feeling. At that stage heavy mittens — known locally as bear-paws — together with violent exercise was the only way to warm the hands.

We flew from Christchurch to McMurdo on 14 October in a US Air Force C-124 Globemaster. Back in Washington there had evidently been some fireworks between NSF and the Navy concerning the delays that we encountered last season. Also targeted was the high priority that the Navy had given to representatives of the Press, generally at the expense of the scientific program. The result was that Admiral Tyree now addressed me as Charles and asked what he could do for us. Helicopter rides for the Press were confined to the late evening; day flights were for scientists. Relations between scientific staff and the support force became more constructive and helpful, although I personally had found few unhelpful people the previous summer. The change, I felt, was splendidly typical of the American way of doing things. The right course of action dawns late but then — as though they meant it all along — they go all the way. The journalists were less pleased.

A McMurdo innovation this year was a nuclear power station under construction half way up Observation Hill. Yet barely a stone's throw away, the charming anachronism of men queuing to get into the bumfreezer, as the outside privy was known.

Sir Vivian Fuchs was visiting McMurdo as a guest of NSF, so we were able to show him our mechanical dog teams. Four years earlier, Fuchs had led the Commonwealth Trans-Antarctic Expedition on the first overland crossing of Antarctica.[1] Wally Herbert, who some years later hit the headlines when he succeeded in driving a dog team across the Arctic Ocean from Alaska to Spitsbergen by way of the North Pole, was working just over the hill at Scott Base. He was so well respected in dog-sledging circles that when the New Zealanders wanted to add a dozen dogs to their stock, they commissioned Wally to select and buy them in the Disko Bay district of Greenland. Wally was a good friend but he never concealed his contempt for our noisy way of sledge travel. Our introduction of motor toboggans the season before had led to endless debate in the New Zealand camp. The argument was summed up in an article in a New Zealand journal headed "Tin dogs versus shaggy dogs."[2]

Taylor instructed Thiel's group in traveling light — a proposition they found alarming — and Rundle assured them that living off sledging rations was not some subtle form of masochism. The National Science Foundation had accepted the advantage of planned diets and now offered three varieties: a lightweight or "Swithinbank" ration — 30 percent heavier than ours; a medium-weight or "Minnesota" ration — more than twice the weight; and a heavy-weight ration for oversnow tractor traverses where weight did not matter. However, all of our party put on body weight during the season — hardly a sign of privation.

Byrd Glacier was first on our list, and after a week of preparations, we flew there on 23 October in two helicopters. John Hickey, who had done so much for us the year before, was pilot of one of them, and Buddy Krebs the other. Hickey flew Thiel and Olson out over the glacier, and after much searching found three places where they could just find space to land between ice ridges. At each stop the geophysicists jumped out to read the gravity meter and a pair of aneroid barometers.

A gravity meter can detect differences as small as one-hundred-millionths of the earth's gravity. In principle it consists of an extremely sensitive spring balance with a

Army helicopter at Mount Hope after landing me for gravity measurements at a dozen points across Beardmore Glacier.

small weight on the end, but in practice there are complications. It is sensitive to temperature and to the varying gravitational attraction of the sun and the moon. Ice being less dense than rock, the force of gravity on a glacier is less than it would be if the glacier were instead made of rock. This makes it possible, with certain assumptions, to make a rough estimate of the depth of ice. The method is critically dependent on knowing the height of all observing stations above sea level — hence the barometers.

Because of the continuous crevassing, there was no question of seismic depth sounding on this glacier. Seismic reflection shooting, which calculates the depth to bedrock by timing the shock wave from a small explosion on the surface, has never been successful in badly crevassed country, and Byrd Glacier was no place for experimentation.

We returned to McMurdo the same evening. I had asked Philip M. (Phil) Smith, the NSF representative, to arrange for a reconnaissance flight to circumnavigate the Ross Ice Shelf — partly to see how, in some future season, we might tackle ice movement measurements along the eastern margin (the Siple Coast), and partly to judge what we might soon find along the southern and western margins south of Beardmore Glacier.

On 21 October the air squadron provided a P2V Neptune named *Bluebird*, a strange half-breed of an aircraft with two piston engines and two jet engines. Before jet engines, the Neptune had been a twin-engined bomber with a kind of greenhouse on its nose to offer an all-round view for the bomb-aimer. The nose was an ideal place for sightseers. I climbed into it and at 2240 we launched into the sky with the help of four engines and twelve JATO rockets. I had asked for a night flight because with long shadows it is possible to see much more of the texture of the snow surface.

The pilot was Lieutenant Elias Stetz, USN, and there were five other crew members including a photographer sent to record anything that we needed for later reference. The machine climbed to 10,000 feet at which point the jet engines were shut down and a partition door magically unfolded across each jet intake. This was the economy cruise mode. We passed over Little America V, then Okuma Bay before turning south towards Siple Coast. It took eight hours to reach the southern end of the ice shelf and the landscape en route was awe-inspiring. Five giant ice streams flowed in from the east, one of them so badly crevassed that not even a helicopter could put down on it. A memorable moment was when my reverie was interrupted by a crewman reaching into the nose cone to hand me a hot roast-beef dinner. It was a far cry from the kind of Antarctic travel to which I was accustomed.

We circled over Supporting Party Mountain, one of a chain of peaks discovered by Laurence Gould's Queen Maud Range Geological Party during Byrd's first expedition in 1929. Turning homewards, we flew along the foot of the Transantarctic Mountains. I spent the time scribbling notes on where — and where not — to travel on the ground. We landed at McMurdo 11½ hours after leaving the night before. I felt euphoric at having learned so much so quickly from the panoramic view afforded by the nose cone. An unbeatable aircraft, I thought, for its reconnaissance role.

On 30 October our combined six-man party drove to the airstrip with a ton of gear behind each of three motor sledges. It took two of the larger R4D-8 "Skytrain" aircraft to swallow everything. One of them had the name *Wilshie Duit* painted on the nose. The Nansen sledges had to be lashed to the ceiling. We blasted off the runway with eight JATO rockets apiece and headed for Nimrod Glacier. As sole passenger in the lead aircraft, I presided over 5,500 pounds of varied belongings and contemplated some of the contrasts of our journey. Now we were heading south at 120 knots to the smooth roar of two big engines. Tomorrow we would continue — at four knots — to the sound of three weary lawn mowers. One could wish for some happy medium.

The pilot, Jim Weeks, made two passes over the chosen landing place and had me lean over his shoulder and swear that the surface patterns that we saw below were sastrugi and not crevasses. Sastrugi are sharp, irregular ridges formed on the snow surface by wind erosion, and unlike crevasses, they are not inherently dangerous. There were crevasses around, but not within a mile of where we landed. One feels a sense of apprehension in choosing a landing place for a big aircraft, for the stakes are high. A series of jarring lurches as the skis touched the snow showed that my estimate of the height of the sastrugi was wrong.

Two weeks after this Jim was at the controls of the same aircraft when it finally ended its days on rough sastrugi, fortunately without casualties. After we came to a stop on the Nimrod he looked unhappy, but radioed the pilot of the other aircraft to land beside us. When the aircraft had been unloaded, both took off with a fiery burst of JATO to avoid further abuse of the landing gear.

We found ourselves exposed to a 30-knot wind with an air temperature of -2° F. The normal camping procedure is that as soon as a tent is pitched, one man makes his way inside by way of the sleeve entrance while his companion hands in everything needed for the night. The outside man then shovels snow onto the tent flap and secures everything in preparation for a blizzard. His final act before seeking the shelter of the tent is to plant the snow shovel beside one of the tent pegs — always the same one — so that in the unlikely event that the tent is carried away, we would know where to look for the shovel. Our newcomers Linder and Olson pitched their tent but, surprisingly, disappeared inside for the night without piling snow blocks on the outside flap. Having experienced only light winds before, they must have assumed that the vigor with which the rest of us piled snow round our tents was just a quaint form of muscle-building. Schooled, as I was, in the hard-won traditions of Antarctic sledging, I came to understand that in a technological age, men brought to the Antarctic can be blissfully unaware of the elementary recipes for survival. Perhaps, for the first time in their lives, they had landed where nobody was going to do things for them. "Typical," quipped Tom, "of the dependent generation." We ought to have proffered advice but guessed that it might be unwelcome. However, they learned quickly and their tent was never blown away.

We had 19 miles to go from the

landing point to our old camp off Cape Lyttelton, and this was covered on 1 November. The sun was shining and the temperature a warm +27° F. The next day we skied across the glacier to Cape Wilson with the gravity meter and a pot of paint. Paint is more viscous at low temperatures but it can still be used. The gravity meter was set down and read at each stake, and fluorescent paint was applied to make the bare aluminum pipes more conspicuous from the hilltop. Being early in the season, the crevasses along the southern margin were well bridged, and we reached the hill near Cape Lyttelton without difficulty. It was a relief to see the stakes in the telescope, and a joy to find them where we had been led by the earlier measurements to expect them. Thiel remained in camp to do seismic work, and by arrangement, he was picked up by an R4D on 4 November. He planned to be away for a week doing airborne magnetometer measurements on a flight to the Soviet Mirny Station and the Australian Wilkes Station. With the Nimrod work completed, I headed with the rest of the party for Beardmore weather station.

The accuracy of our navigation was much improved by large liquid-filled compasses of the type used in Hurricane and Spitfire fighters in World War II. I

Out of sight of the mountains, or in fog, we used the sledge wheel to measure distance traveled. A compass from a World War II fighter plane is mounted on the left rear corner of the sledge beside the driver. Tom Taylor carries alpine rope, ice axe, and crampons to cover the remote chance that the whole train might fall into a crevasse.

had bought them in a war surplus store in London. We mounted one on the rear sledge so that the helmsman of each train could maintain a steady heading in any weather. Gravity work requires a continuous record of position; compass and sledge wheel were needed for dead-reckoning between stations fixed by sun sights. One day was spent establishing the position of six ice movement stakes that we had planted on the ice shelf the previous season. All had moved in a northerly direction, the fastest at the rate of 1,380 feet per year.

It was a terrible shock to each one of us to be told, during our routine radio contact with McMurdo Station on 10 November, that the P2V Neptune *Bluebird* in which I had flown two weeks earlier had crashed at Wilkes Station. Shortly after take-off there was an explosion followed by a fuel fire. Of the nine men on board, Ed Thiel and four of the aircrew were consumed in a ball of fire. Elias Stetz, the pilot on my ice shelf flight, was one of the survivors.[3]

We lost not only a dear friend but also the mainstay of our geophysical programme. Ed had a lovely young wife at home — now a widow — and we shed tears thinking of her predicament. We hardly knew the four who died with him, but their families and friends too must have been devastated.

Olson and Linder bravely decided to carry on as best they could. Our gravity

The P2V Neptune in which Ed Thiel and four of the aircrew perished.

Tom Taylor (right) driving the Polaris by remote control from 60 feet behind. Crossing Beardmore Glacier we felt relatively safe with this method even in badly-crevassed areas.

work continued for the rest of the season, but because of instrument troubles, there were few results from the seismic ice-depth sounding. It was not a happy group that sledged into Beardmore Station seven days after departing from Nimrod Glacier.

After being confined to the tent by blizzards for two days, Taylor, Rundle, and I got under way for Beardmore Glacier. It was 17 November, my 35th birthday. As we had nothing to help a celebration, I did not tell the others. This time we chose to carry all our gear on a single train behind the Polaris motor sledge. To guard against the remote possibility that the whole train might disappear into a crevasse, leaving us on the surface with no means of survival, the driver wore a rucksack containing alpine rope, crampons, and an ice axe. The plan was that one of us would be lowered into the crevasse to send up the gear, one box at a time.

Rundle had a moment of terror. A snow-bridge gave way under the full length of one of his skis, and in that moment he glimpsed the inky blackness below. But he recovered his balance and carried on, albeit with a paler complexion. After 28 miles we camped below the Mount Kyffin icefall within sight of one of the movement stakes. It was good to know that the stakes had survived the winter, for we were afraid that they might have been blown over or snowed under. After digging a pit to measure the density

Camp in the Thiel Mountains, named for our colleague Edward C. Thiel.

of the snow, we crossed the glacier, painting each stake and reading the gravity meter as we had done on the Nimrod. Rundle built snow cairns to make the stakes more conspicuous. He built two of them without taking his skis off, because in some places we were never quite sure what we were standing on or digging into. One glaciologist was known to have disappeared down a hole of his own making — through a snow bridge.

Less used to skis than the rest of us, Rundle sat astride one of the cargo sledges crossing the worst of the crevassed area. A compulsive smoker, at one stop he jumped off and began to pace up and down with a cigarette, blissfully unaware of the snow bridges all around. My heart almost stopped when I saw it. To teach him a lesson, we broke open one of the crevasses with a ski pole and made him lie down to peer into the black depths. It was a sobering experience. My diary records: "Arthur nearly stepped into his grave twice despite all warnings."

Air temperatures in the first half of November hovered around 0° F. The lowest recorded was at midnight on 11 November, -15° F, though there must have been colder moments while we were asleep in the tent. On 20 November we climbed to the old survey stations on Mount Hope and found all twelve stakes in the telescope. In camp that evening, we were able to end half a century of speculation by calculating that the glacier had moved at the rate of 42 inches per day — 1,278 feet per year — since our last visit. Later I was able to show that this rate remains the same during both summer and winter seasons.

Next day we struck camp and crossed, for the sixth and last time, our Beardmore "Valley of the Shadow." None of us was sorry to leave the area. Reading the gravity meter every three miles, we worked our way back to the weather station.

The silence was broken on 24 November by the arrival of a small air armada supporting a trilateration traverse along the mountain range in aid of mapping. Trilateration — measuring distances by radar — had largely superseded triangulation as a method of determining positions. While they were preparing to start, I hitched a ride with their supporting Otter almost to the head of the Beardmore Glacier. The pilot was Lieutenant Ronald (Ron) Bolt, USN. It was a clear day, and we climbed up the middle, past all the landmarks made famous first by Shackleton and three years later by Scott — Mount Hope, the Granite Pillars, Socks Glacier, The Cloudmaker, Buckley Island, and the Dominion Range. At the point where Mill Glacier joins the Beardmore we saw a huge expanse of absolutely smooth crevasse-free and snow-free blue ice. I wondered why nobody had used it for landing wheeled aircraft. Afterwards I described the area to the aviation fraternity at McMurdo. So foreign was the idea of using wheeled aircraft in the interior of the continent that it took twenty seven years before one actually put down on the patch of ice that we had found. When finally it did, I was in the copilot's seat.

On 25 November I was able to take advantage of a new US Army Bell UH-1A turbine helicopter to complete our gravity section across the glacier. Whereas the Navy helicopters had a two or three-man aircrew, the Army preferred a one-man aircrew and for this reason were able to carry a greater payload. The pilot, Lieutenant Greene, was a Korean War veteran who did not waste time. I occupied the copilot's seat; behind me were Tom Taylor and Bill Chapman with tellurometers. The tellurometer is a precise electronic distance-measuring instrument that needs to be placed at each end of the survey line. We flew to the summits of Mount Hope and Mount Kyffin, and compared these armchair ascents at 2,000 feet per minute with our normal perspiring climbs at 1,000 feet per hour. Evidently we were born just a few years too early.

On Mount Kyffin, a sharp peak, there was no place to land the helicopter, so Greene hovered with one skid on a ledge and the other in mid-air while the surveyors got out. I found the experience hair-raising but was told that such maneuvers were all in a day's work. Landing for gravity measurements at a series of points across the glacier, Greene jumped out and carried my instruments to where I needed them as courteously as any hotel porter. A pity, I thought, that there were no Navy pilots watching.

One day John Hickey, with his helicopter and my connivance, broke the rules by leaving me alone on Mount Hope to survey our Beardmore stakes. It is normally plain common sense, not to mention an NSF standing order, that nobody should be left alone away from camp — or in camp for that matter. But we were short of available helpers that day, so we accepted the slight risk. I took a sleeping bag and some food, assuring Hickey that if for any reason he had trouble returning, I would be safe and well for 10 days or so.

He returned to pick me up at midnight.

[1] Sir Vivian Fuchs and Sir Edmund Hillary. *The Crossing of Antarctica.* London, Cassell & Co., 1958.

[2] Peter Otway. Tin dogs versus shaggy dogs. *Antarctic* (quarterly journal of the New Zealand Antarctic Society), Vol. 3, No. 6, June 1963, pp. 232–235.

[3] David Burke. *Moments of Terror. The Story of Antarctic Aviation.* Kensington, New South Wales Press, 1994 (p. 273).

Chapter Seven

The Far South

*It falls to the lot of few men to view land
not previously seen by human eyes.*
　　　　　　　E. H. Shackleton, 26 November 1908

After a two-week delay made necessary by the commitment of all large aircraft at McMurdo Station to other tasks, we were picked up by an R4D on 7 December. Into the cabin we crammed two motor sledges, four Nansen sledges, 40 aluminum pipes, communications and time-signal radios, surveying instruments, camp gear for three men, motor fuel for 500 miles, food, and cooker fuel for six weeks. Olson and Linder remained at Beardmore Station to await some spare parts for their seismograph. Flying along the mountain front, we found that the easternmost glacier, named after Scott, was under cloud. The pilot, Jim Weeks again — with a new aircraft — could not see the snow surface and so could not land. He eventually put down in the nearest patch of sunshine. It was on the ice shelf 30 miles off the mouth of Amundsen Glacier and 80 miles from Mount Hamilton, our intended destination on Scott Glacier.

If there was some disappointment at the need for an extra sledge journey, we soon forgot it. Spread out before us was an unbroken sweep of 200 miles of the Queen Maud Mountains from Mount Wade in the west to Mount Gould and beyond in the east. We could recognize most major features of the landscape from the 1931 reconnaissance map of the first Byrd expedition. While preparing for the season, a copy of the map had been given to me by Captain Ashley McKinley, Byrd's aerial photographer. Now, 30 years after it was published, it was still the only map of the area, so we plotted our tracks on it. McKinley had penciled notes on my copy like "Mt. Gould about here," and in two places, "Delineation not correct."

Fifty years earlier, and 18 years before Byrd's expedition, the Norwegian explorer Roald Amundsen was the first to cast eyes on this scene while he and his four companions were heading for the South Pole.[1] From our camp we looked straight up Axel Heiberg Glacier and tried to pick out the route that Amundsen had followed on his climb in November 1911. Byrd flew down the same glacier on returning from his epic flight to the South Pole on 29

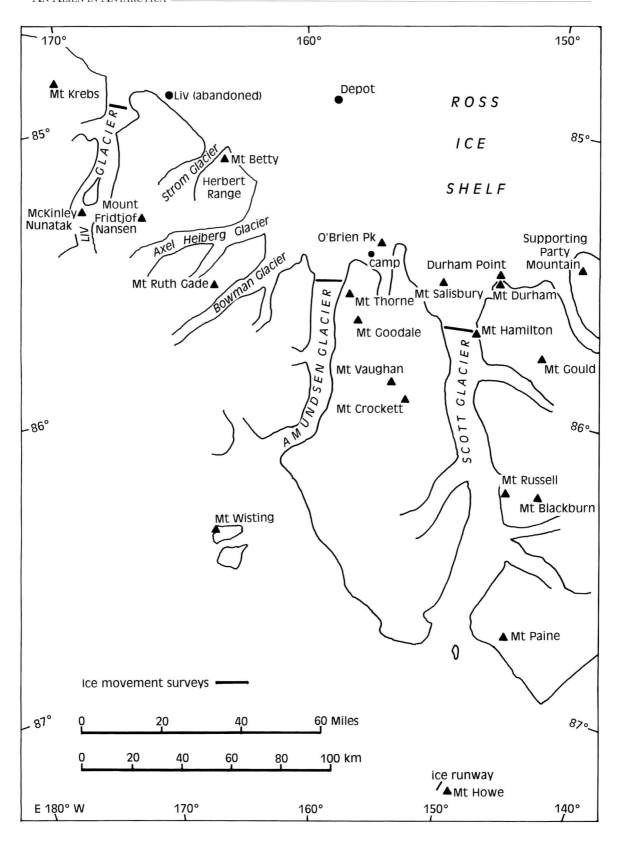

November 1929. Byrd's geologist Larry Gould, who was stationed at the foot of the mountains to provide weather observations for the polar flight, later agonized "Here was Commander Byrd and his party covering in four hours a distance which took us four weeks to cover with our dogs."[2]

The mountains themselves were on a massive scale: Mount Fridtjof Nansen, a glistening 13,000-foot snow dome, contrasted with the soaring rock ridges of the main escarpment.

After sorting gear and making a cache, we got under way. With only three men to drive two motor-sledge trains, someone had to go solo, and this was quite exciting until we got used to it. Solo driving requires more competence on skis, because slow recovery from a fall could mean a runaway sledge train. It would be embarrassing to see one's worldly goods disappearing over the horizon. We soon learned to allow a long rope to trail behind the second sledge, so that the driver had time to struggle to his feet, grab the rope and haul himself back to the steering position. Rundle was still not confident on skis, so Taylor and I shared the solo driving. Though there were moments when our machine seemed bent on escape, it never quite got away. We covered the 65 miles to Durham Point at the mouth of Scott Glacier in 2½ days. This journey was particularly trying for the motor sledges, as over much of the distance the surface consisted of rough, hard sastrugi up to three feet high. There were capsizes, but both vehicles survived the ordeal without visible damage. The last few miles into the

Liv Glacier to Scott Glacier.

land were over bare blue ice.

At Durham Point we came upon a small rock cairn. Inside was a note reading:

Thanksgiving Day, Nov.29/34, Thorne Glacier. This note indicates that last night we of the Queen Maud Geological Party of the Byrd Antarctic Exp.II camped near this point en route up the Thorne Glacier on geologic reconnaissance. In honor of the day we name this topographic feature "Durham Point" after Durham, New Hampshire. Stuart D. Paine, Richard Russel [Russell], *Quin A. Blackburn. Nov.29, 1934.*

Beside the cairn was a stadia rod, plane-table tripod, two ice axes, and two pairs of crampons. By the time they returned to base at the Bay of Whales on 11 January 1935, the party had covered 1,410 miles with their two dog teams.[3] Thorne Glacier was later renamed Scott Glacier (for Robert Falcon Scott) by a decision of the US Board on Geographic Names. Thorne, a member of the Byrd Antarctic Expedition of 1928–1930, had a mountain named after him.

We climbed Mount Durham to have a look across the glacier, and were dismayed by what we saw. Not only was it wider at this point than the Beardmore, but it was too scarred by crevasses for us to attempt a crossing. We would have to climb higher up the glacier in the hope of finding a narrower and smoother part. So we loaded what we needed behind the Polaris and started south the following day. My diary reports:

A hell of a day. Drove 17 miles and 1250 feet up glacier, most of it on bare ice. The Polaris took a tremendous beating on the sun-cupped surface. Got bogged in a crevasse area and had to back out, me driving as the holes were a few feet apart and we had to go between. On level patches we sat astride [the sleds] like human torpedoes. Much of the time we all wore crampons, getting off the sledges to push on every upslope. But on several snowy stretches we had to stop to take off crampons and put on skis, as this was safest in the rather unpleasantly crevassed surface. Arrived at last at Mt Hamilton, finding a snow patch to put the tent on inside the lateral moraine.

Here the glacier was narrower and there was a possibility of getting across to plant stakes. We were only 255 miles from the South Pole. I thought of chugging in to the pole station just to see the expression on their faces; we could have done it with the seven jerrycans of fuel remaining.

Scott Glacier (looking upstream) from an altitude of 24,000 feet. We camped beside Mount Hamilton (left) where the glacier is eight miles wide. The South Pole is 200 miles beyond the far horizon.

Leaving Mount Hamilton to cross Scott Glacier. We had to abandon the attempt in a maze of crevasses.

Each evening as the sun got lower in the sky we were subjected to loud cracking noises in the ice, almost like an artillery barrage. I think the surface of the ice was contracting after the heat of the day and fracturing in the process.

Throughout our 12-day stay at Mount Hamilton the wind blew incessantly, apparently caused by the channeling of cold air down the valley from the polar plateau — the so-called katabatic wind. It made a wearisome task of every bit of work outside the tent. Tom tried to set up the theodolite, but it shook too much to be used.

The 1934 party had the same problem in the same place. Russell told Byrd later that if they ever wrote a book about the journey it would be called Wind. Byrd put it like this:

> *The wind had bitten into their memories as acid bites into an engraving plate…Up the glacier, the mountain walls of which make a funnel, the wind was even worse. It buffeted them like a solid force. It never blew less than 20–40 miles per hour, and at times approached hurricane force.*[4]

On 15 December we loaded a single Nansen sledge and started across the ice, man-hauling. Crevasses began after half a mile. They got bigger and bigger. Rundle and I were in the lead, jumping the smaller crevasses and going round the big ones. With a 600-pound load behind us, we were afraid of being caught off balance as the traces jerked taught. The crevasses

were so close together that if we had slipped, one of us might have gone in. Taylor gallantly man-handled the back of the sledge to prevent it from slipping sideways into other hazards. On one occasion the full 13-foot length of one sled runner skidded on to a frail snow bridge, and we had visions of being dragged backwards by our traces. After three hours we were a third of the way across the glacier but in a veritable maze. We were pushing our luck. Reluctantly, I turned back, planting four stakes and reading the gravity meter on the way in to Mount Hamilton. In camp that night we drank a toast (in medicinal brandy) to Roald Amundsen and his gallant men, who reached the South Pole exactly 50 years before.

We sweated up Mount Hamilton to a height of 2,750 feet above the camp, carrying the theodolite and two tripods. Looking across the glacier, it was a relief to find that there were a number of moraine and sharp ice features on its surface well beyond our farthest stake, so we were able to select four of these to serve as additional movement markers. The Beardmore measurements had shown that we could safely extrapolate the amount of ice movement from observations taken only a few days apart. I therefore decided to wait at Mount Hamilton until each marker had moved through an angle that was measurable with sufficient accuracy to make a second visit unnecessary. After the first set of observations we spent a week on routine baseline measurements and survey work for mapping purposes. The news came over the radio that Olson and Linder had been landed beside Amundsen Glacier and were busy doing seismic soundings.

Considering that our camp was 2,100 feet above sea level and a stone's throw from the South Pole, the air temperatures encountered were remarkably warm. Day temperatures were consistently around +20° F, although wind chill made it feel like -10° F. Rundle drilled into the ice, measuring temperatures every three feet. It was a surprise to find that the temperature 33 feet down in the ice was only -1° F, indicating that our warm midsummer weather was quite normal for the area. The average mean annual temperature for this latitude elsewhere in Antarctica is probably more like -20° F to -30° F.

At an altitude of 4,000 feet on top of Mount Hamilton we experienced the only complete calm of our two-week stay on Scott Glacier. We could see more than 100 miles in every direction, and looked down on a wide basin that extended far into unexplored country beyond Mount Gould. The glacier itself appeared to meander through the mountains, and from this height its tortured surface looked as peaceful as the rippled waters of a stream. After a final set of angles on 22 December, we started for Durham Point. The middle of the glacier had moved a thoroughly measurable 25 inches a day during our seven-day period of observation.

Picking up the Eliason and our other possessions, we headed for Amundsen Glacier, making a 30-mile detour on to the ice shelf to avoid extensive bare blue

Tom Taylor using the theodolite on the summit of Mount Hamilton while Arthur Rundle records the numbers.

ice areas that we had seen on aerial photographs. It was rewarding to find a series of active tidal strand cracks — tiny crevasses with upturned edges — only 11 miles from Mount Durham, for they showed beyond doubt that even at this point some 500 miles from its seaward margin, the ice shelf is afloat, and rises and falls with the tides. There was no earlier record of strand cracks being seen so far south, nor so far from any ice front.

We covered the 90 miles to the foot of Amundsen Glacier in 3½ days. The motor sledges were cruelly treated by sastrugi, and one day the front axle of the Polaris would have fallen off if Rundle had not spotted the trouble in time. We quickly made repairs. Christmas was celebrated on 26 December with our usual pemmican stew washed down with an ounce of whisky. Climbing a steep slope beside O'Brien Peak, we came over a ridge to find the seismic party camped on an unnamed tributary of the Amundsen. Our joy at finding them was redoubled when they handed us the December mail. They had made a successful sounding indicating an ice depth of 2,500 feet and were waiting — wisely — for company before proceeding on to the main stream.

The next day we left them in camp and headed for the glacier to reconnoiter a way. Optimistically, we took with us 10 stakes, the gravity meter and camp gear on two sledges behind the Polaris. But after only seven miles, in which we several times drove across broad sagging snow bridges, I called a halt. Crevasses were on every side, trending in various directions, and some snow bridges had evidently collapsed quite recently. We had to reverse the machine and each sledge in its tracks, because there was no room even for a tight turn without running into danger. Once more frustrated, we retreated to the seismic camp. The weather kept us there for the next 12 days. Seven of these were so completely overcast that there were no shadows of the snow surface. Under this kind of whiteout condition, it is quite possible to drive straight into an open crevasse without seeing it.

For five days we were confined by high winds, occasionally of hurricane force. The roar of flapping canvas and the whistle of wind over tent guys drowned all conversation. Sometimes we feared for the tents.

My definition of a hurricane is a wind of such force that it feels safer to put on every layer of inner and outer clothing, windproofs, and even boots in case the tent carries away. We then wriggle into the sleeping bag and zip it closed. If the tent does disappear, we plan to stay in the bag in the open, even though we would be snowed under. The alternative — trying to make a shelter while being blown off one's feet or succumbing to hypothermia — is riskier.

I have always found that a tent in a blizzard is the best place to read the classic narratives of the heroic age of Antarctic exploration. The stories come to life in a way that they never do in an armchair back home. Amundsen, Scott, Shackleton, Mawson, Byrd — I have read all of them in flapping tents. We read while prone in a sleeping bag. The only drawback is that one's arms get cold because they have to be outside the bag. Sometimes we had enough Coleman fuel to keep the stove running at low heat; at other times we gave up and slept.

Understandably, none of the party felt

elated on establishing a new Antarctic record for physical inactivity. Tom said the place should be called Foggy Bottom. But since that would never be approved by the US Board on Geographic Names, we might call it Sun Valley. Only we would ever know the secret. We read and then exchanged books. Once or twice we went

schedule, Beardmore Station had not answered since 23 December. The five-watt output of our transmitter may have had something to do with it. Finally on 3 January 1962 we heard an R4D flying along the mountain front in the clouds, and guessed that people at McMurdo had become anxious enough to search for us. I

The 1961–1962 field party fogbound in camp on Amundsen Glacier.
From left: Jim Olson, Art Rundle, Hal Linder, Tom Taylor, and the author.

outside to play cricket, Rundle serving as coach. The bat was a shovel, the balls were snow.

Tom and I never got tired of the food, but the folks in the next tent suggested that our meat bars were made from dogs run over in the backstreets of Chicago by the employees of the meat packing company.

Although we called daily on our radio

called on the radio but, as we afterwards learned, they were too busy looking for us to listen as well.

Two days later, on our eighth consecutive day in camp, we were wakened at 0200 by an aircraft banking overhead. They were flying between low overcast clouds and the invisible snow, but they had seen us. They let-down in total whiteout, banged on to the surface and taxied up to the camp. Out jumped Phil Smith and pilot Ron Carlson, an old hand at landing big aircraft in strange places. Phil said it was their fourth attempt to find us. They brought mail and some food, and agreed to ask the South Pole Station to keep radio watch for us during the rest of the season.

One of the aircrew, accustomed to remote landings in support of the New Zealand expedition's dog teams, looked round and asked where our dogs were. Taylor pointed at a motor sledge and said with a straight face, "Over there!" The man was taken aback. Like Wally Herbert, he felt that our ugly machine was a poor substitute for a family of tail-wagging huskies.

After satisfying themselves that all was in order, our visitors blasted into the clouds with seven JATO bottles. For our little party, more than 500 miles from base, it was strangely moving to think that six men had given up two nights' sleep and flown thousands of miles, just to find us and to see if we might be in need of assistance.

True to their word, South Pole radio came through loud and clear the following day. They heard us, and we had no more trouble with communications. The weather cleared on 10 January and we struck camp. Rundle, Olson, and Linder went down to the ice shelf for seismic and gravity work, while Taylor and I, with the survey gear on a single sledge behind the Polaris, drove 10 miles up-glacier and camped at a point 2,000 feet above sea level. Here we could overlook the glacier where it appeared to be about 10 miles wide. As with Scott Glacier, there were plenty of crevasses, pinnacles, and rocks being carried along by the ice to serve as ice movement markers, so we had no difficulty in measuring the rate of movement over the next five days. When the usual observations were complete, we sped down to the ice shelf and rejoined the rest of the party at the 7 December cache.

One of the leftover four-gallon metal jerrycans, full to the brim with cooking fuel when we left, was unaccountably empty. We looked carefully for any hole in the can, but found none. Fortunately the loss was not serious, as we carried some fuel in reserve. But I recalled the similar

Ron Carlson blasts off from whiteout camp off Amundsen Glacier.

experiences of Scott and his companions on their return from the South Pole in 1912. Scott's diary records unaccountable fuel losses from three separate depots on the ice shelf. Our explanation accords with that of Leonard Huxley, who edited Scott's diary for publication:

> *As to the cause of the shortage, the tins of [kerosene] oil at the depôts had been exposed to extreme conditions of heat and cold. The oil was specially volatile, and in the warmth of the sun . . . tended to become vapour and escape through the stoppers even without damage to the tins . . . However carefully re-stoppered, they were still liable to the unexpected evaporation and leakage already described. Hence, without any manner of doubt, the shortage which struck the Southern Party so hard.*[5]

Olson and Linder had found the ice thickness to be 2,950 feet with 1,100 feet of water beneath it, confirming the existence of a deep trough in the sea bed near the mountain front. Here was the thickest ice yet found on the Ross Ice Shelf. Crary later judged their records " . . . as good as any I have seen from the Antarctic."[6] This was a fine tribute to our two young seismologists — now bereft of their leader. Dimensions like these never ceased to amaze us. We were insignificant creatures in a land of immense proportions.

Rundle had drilled into the ice to find that the mean annual temperature hereabouts was -5° F. On 12 January he recorded an air temperature of +33° F. We decided not to report this in letters sent home because it could explode the myth of eternally numbing cold associated with high latitudes.

The next day we headed for Mount Betty, a nunatak named by Amundsen for his family's housekeeper. A nunatak is a small mountain surrounded by a glacier. The Norwegian *fjell* evidently suffers in translation to the English mount, for Mount Betty was at most 100 feet from top to bottom. But on it were two big cairns, one built by Amundsen's party on their return from the Pole in 1912, the other built by an exploring party of the first Byrd expedition in 1929.[7] The leader of the American party was Laurence (Larry) Gould, then Assistant Professor of Geology at the University of Michigan. So we were not the first people to make the long voyage from Ann Arbor to the Queen Maud Mountains. Gould's cairn contained a note that read:

> *Dec 25, 1929. On this day at 1030 p.m. the geological party of the Byrd Antarctic Expedition stopped at what we believe to be Mt Betty on our way back to the west side of Axel Heiberg Glacier from a 100 mile trip eastward along the foot of Queen Maud Range. Here we discovered a cache laid down by Amundsen on his way northward from the Pole. We have carefully replaced all rocks in cairn and leave it intact as we found it except for the bit of writing which Amundsen left in a tin can.*[8] *The members of the geological party are L. M. Gould, University of Michigan, Ann Arbor, Michigan, USA; N. D. Vaughan, Hamilton, Mass, USA; E. E. Goodale, Ipswich, Mass, USA; F. E. Crockett, Ipswich, Mass, USA; G. A. Thorne, 1130 Lakeshore Drive, Chicago, Illinois; J. S. O'Brien. L. M. Gould, Leader.*

When Gould's party returned to

Tom Taylor beside the cairn built by the Norwegian explorer Roald Amundsen on returning from the South Pole in January 1912. It contained an intact can of kerosene which we would have used in an emergency.

Mount Betty a few days later, the men built their own cairn in which they left this note:

> *December 28th 1929, 5 a.m. On this date at the above time, the Geological Party of the Byrd expedition returned to Mt Betty after having called here Christmas morning where we discovered a cache laid by Amundsen 18 years ago. We have placed here in a cairn some equipment which we are discarding as we turn northward toward Little America. We have been gone from Little America since Nov. 4 and have in the intervening time made a rough chart of 150 miles of the Queen Maud Range and I have made geological studies which demonstrate that this range continues even to 145 west longitude as a major feature of the landscape — just as it is far away on the western side of the Ross Sea. We expect to leave tomorrow morning headed northward toward Little America with 21 dogs and two months supply of man food. The members of the Geological Party are L. M. Gould, geologist and leader; J. S. O'Brien, surveyor; G. A. Thorne, topographer; N. D. Vaughan, dog driver; F. E. Crockett, dog driver and radio operator; E. E. Goodale, dog driver; by L. M. Gould.*[9]

A broken Nansen sledge and a camera tripod were lying against the outside of the cairn. Inside was an assortment of clothing, some first aid supplies, radio parts, dog harness and Gould's geological hammer. Knowing that Gould was still very much alive, I took the hammer and afterwards mailed it to him at the University of Arizona with these words:

> *Here is the hammer that you left at Mount Betty some time ago. I hope that its absence has not caused any great inconvenience during the intervening [32] years.*

Fifty yards from Gould's cairn, Amundsen's contained a further note from Gould and a four-gallon can of kerosene. The Norwegians, wisely, had soldered the lid on the can; now, after 50 years of exposure, it was still full. An added irony of our visit was that it coincided with the 50th anniversary of the day that Scott's ill-fated party reached the South Pole.

Gould's party of 1929 was traveling in a different age from ours, when American policy was more overtly political. At the easternmost point of his journey, he had left a note with the words:

> *We ... claim this land as a part of Marie Byrd Land, a possession of the United States of America. We are not only the first Americans but the first individuals of any nationality to set foot on American soil in the Antarctic.*[10]

It was Byrd's intention to claim sovereignty for the United States of the sector of Antarctica between longitudes 80° W and 150° W. The limits were set by the United Kingdom's claim, dating from 1908, to a sector east of 80° W, and New Zealand's claim, dating from 1923, to a sector west of 150° W. At Gould's farthest east he was at 147° 33' W, just within "the American sector." The State Department did not pursue Gould's claim and that sector remains, to this day, formally unclaimed. It is now too late because Article IV of the Antarctic Treaty precludes further claims.

I was relieved that my own party had no political objectives. Leaving Mount Betty, we had a rather exhausting climb to get above an icefall on Strom Glacier where there appeared to be a good route to the west. The motor sledges kept breaking traction and digging in, and repeatedly we had to man-handle them, zig-zagging slowly to gain height. At the top there was smooth going on soft snow, and Taylor and I got ahead of the others. At some point we must have passed by the site of Gould's "Strom Camp." On stopping soon after midnight, we had covered 95 miles in two days. After a night's sleep we rattled on for nine miles before coming upon some crooked radio masts and the snowed-under remains of a weather-beaten Jamesway hut. It was the abandoned Liv Glacier weather station of Operation Deep Freeze II (1956–1957). This had been built close to the site of Byrd's aircraft fuel cache for the 1929 South Pole flight. After digging down to the door, we found the hut partly filled with snow but otherwise in good condition. Some left-over five-year-old food served as a welcome addition to our dwindling supplies.

The next day, 18 January, we started for Liv Glacier 15 miles away. After five miles we ran on to bare ice riddled with cryoconite holes. These are formed where windblown sand or fine pebbles, absorbing more of the sun's radiation than the ice, melt themselves downwards. The debris can be seen some way down through crystal-clear ice. In a few of the holes there was water, and a form of algae which, but for a few crustose lichens, is surely the most tenacious form of life on Earth.

Gould had tried to climb the glacier, so we knew what to expect. He found it so steep that he was " . . . scarcely able to climb it on foot even when shod with crampons."[11] To get to a narrow and traversable part of the glacier we had to by-pass an icefall. This meant climbing a very steep snow slope up a narrow apron between the rock walls of the valley and the first crevasses. By using alpine ropes to connect two motor sledges in tandem to a single Nansen sledge, the motor sledges were beyond the steepest part of the slope by the time the towed sledge was climbing it. At the top of the slope we made camp in a windscoop at the foot of an unnamed nunatak.

As soon as the weather was favorable we crossed the glacier on skis, finding it to be only six miles wide, and as such the narrowest that we had worked on. We reached the far side without difficulty, but being late in the season, the crevasse bridges were treacherously weak, and we decided that trying to reach the middle of the glacier with the seismic gear would be too great a risk. However, we did succeed in making the best gravity traverse of the season. Seven stakes were set up on the way back to camp.

At the same time as we were working along the foot of the Queen Maud Mountains, Wally Herbert was leading a New Zealand survey party on the polar plateau side of the same range. We spoke on the radio. While we ourselves felt free to travel anywhere in pursuit of the task in hand, with nobody looking over our shoulders, for Wally it was not so easy. He proposed to round off an eminently successful season of mapping by sledging to the South Pole, where he could be picked up by an LC-130 and returned to McMurdo. This was in fact his safest option. However, on contacting the leader of the New Zealand expedition at Scott Base, permission was refused. Wally next proposed descending Axel Heiberg Glacier (the route used by Amundsen in 1911) so that an R4D could pick up him and his party from the ice shelf. Wally takes up the story:

> *How Amundsen would have laughed if he could have seen us, beating our knees with frustration beside a radio set, listening to garbled reports of conferences held between our superiors on whether or not to give us permission to descend Axel Heiberg Glacier. . . . No longer are explorers permitted to face their dangers alone. Nor are they permitted to go off the air and take their calculated risks. . . . What a relief of responsibility it would be for the modern explorer if he had only the safety of his companions and himself to worry about!*[12]

I knew what a privilege it had been to travel with only my party's safety to worry about. It has ever after struck me as absurd that officials hundreds of miles away, most of them with no relevant experience, can have the audacity to pronounce on issues of safety in the field. For Wally, with his vast experience of safe Antarctic travel, it must have been particularly galling.

At this point our party had to be

Crossing Liv Glacier to plant ice movement stakes. Art Rundle (left) and Hal Linder (right) with the three-inch-diameter irrigation pipe. Tom Taylor carries the ice drill and a climbing rope.

divided once more. It had taken two R4D loads to bring us south, and I had requested two flights for the return to McMurdo. Rundle, Olson, and Linder therefore drove down to the Liv Station hut, where they were picked up on 23 January. On his way home the pilot diverted to drop mail. As he flew over our camp, out dropped a little brown bag. The plane circled once to ensure that the bag was not languishing at the bottom of some crevasse, then climbed for the long haul back to McMurdo. Taylor and I devoured the mail and a heap of fresh magazines. "A good thing," my diary notes, "because I was about out of reading matter." We stayed six days longer to complete our observations on the glacier. It turned out to be the slowest moving stream of all; the middle stake was creeping a mere 12 inches a day.[13]

We raced down to the ice piedmont at the foot of the mountains, wearing through many a loop of hemp rope that had to be used to brake the sledges. Running with crampons on one's boots becomes tiring, and it was with considerable relief that we reached the snow at the foot of the slope and put on skis. It was blowing a fresh gale by the time we reached the old Liv Jamesway hut. We managed to pitch the tent after a struggle to keep it from being carried away by the wind. The blizzard kept us inside for four days until an R4D, finding the camp by means of radar, dropped through a hole in the haze and landed beside us.

Back at McMurdo Station, the airstrip

LC-130 Hercules preparing to fly to New Zealand from the sea-ice runway at McMurdo.

was hidden beneath an autumn snowstorm, and our pilot, Lieutenant-Commander Michael Jarina, USN, had to make a ground-controlled instrument approach. After unloading, we crowded into the Navy mess hall to enjoy our first professionally cooked meal in three months.

A large area of sea-ice off Hut Point broke up and drifted away to sea a few days later, taking with it a retired R4D *Marlene*.

There remained only Mulock Glacier to be visited for its second set of measurements, and the five of us were taken there in two helicopters on 10 February. Taylor and I camped on the cliff-top and spent the next three days adding to our earlier survey work and searching for the stakes.

First we followed with the telescope as one of the helicopters carried Rundle, Olson, and Linder over the ice to make gravity readings. They stopped at seven places and then made camp on the only relatively crevasse-free part of the glacier.

Our theodolite had to be propped against vicious gusts of an unceasing plateau wind, clear evidence that the summer season was drawing to a close. My diary reports that with hands made clumsy by cold "I lost the roll of toilet paper to a wild gust." It unravelled as it went. Tom picked up the diminished remains down the hill. Snow is an unsatisfactory substitute.

It was on 13 February, as the helicopters flew in to pick us up, that at last we found eight of the nine movement-stakes and fixed their positions. Finally back at McMurdo Station, we had good reason to thank the Navy for the cheerful, willing and generous support that we had received since October. The air squadron calculated that we had used a total of 22 hours of helicopter flying time and a good many more of R4D time. Taylor and I flew to New Zealand a few days later. Rundle stayed two weeks to put the equipment in storage for another field season.

We left, as most people leave, thinking of all that remains to be done. But we had accomplished a great deal. In all, we made successful velocity measurements on seven of the eight major glaciers in 650 miles of mountain range. We found that Byrd Glacier is the largest and fastest-moving, and Liv Glacier the smallest and slowest. Owing to the tragic loss of Ed Thiel, the seismic work failed to give enough ice depths to calculate the volume of ice transported. But our 140 gravity stations would give an approximation of the cross section of each glacier. In the course of the season, one motor sledge covered a total of 723 miles hauling an average load of 1,500 pounds, while two others each covered around 575 miles.[14] Taylor determined positions by sun or star observations at each glacier visited. By intersecting all visible peaks from several baselines, he brought home a reconnaissance triangulation network covering 150 miles of the Queen Maud Mountains. Rundle had made weather observations throughout the season.

The seven glaciers visited have a combined width of about 70 miles. They move at an average speed of about 1,100 feet/year,[15] which must add some 15 square miles to the ice shelf. If the average thickness is 2,000 feet, and this is quite possible, then six cubic miles of ice are fed into the Ross Ice Shelf each year.[16]

We never really got used to the prodigious scale of the landscape in which we lived.

---

[1] Roald Amundsen. *The South Pole, an Account of the Norwegian Antarctic Expedition in the "Fram", 1910–1912.* London, John Murray, 1912 (2 vols).

[2] Richard Evelyn Byrd. *Little America. Aerial Exploration in the Antarctic, the Flight to the South Pole.* New York, G. P. Putnam's Sons, 1930 (p. 398).

[3] Quin A. Blackburn. The Thorne Glacier section of the Queen Maud Mountains. *Geographical Review,* Vol. 27, No. 4, 1937, (p. 598).

[4] Richard Evelyn Byrd. Discovery. *The Story of the Second Byrd Antarctic Expedition.* New York, G. P. Putnam's Sons, 1935 (p. 363).

[5] Leonard Huxley, (ed.). *Scott's Last Expedition.* London, Smith Elder & Co., 1913, Vol. 1, Appendix (p. 631).

[6] Letter from A. P. Crary to Glenn E. Bowie, 16 October 1963.

[7] Laurence McKinley Gould. *Cold. The Record of an Antarctic Sledge Journey.* New York, Brewer, Warren & Putnam, 1931.

[8] Some years later I came across a reproduction of Amundsen's note in Gould's *Cold* (New York, Brewer Warren & Putnam, 1931, p. 213). My own translation from the Norwegian reads:

> *6–7 Jan.1912*
> *Reached and circled the pole 14th-16th December 1911. Discovered that Victoria Land and — probably — King Edward VII Land are connected at 86° S latitude. Moreover these lands continue in a great mountain range toward the south-east. Have observed this range extending to 88° S. As far as we can judge, it appears to continue in the same direction over the Antarctic continent. Passed this place on the return journey with provisions for 60 days, 2 sledges, 11 dogs. All well.*
>
> *Roald Amundsen*

[9] The originals of both of Gould's notes and also the Durham Point note are in the Archives of the Scott Polar Research Institute, Cambridge, England. Photocopies are in the National Archives in Washington, DC.

[10] Richard Evelyn Byrd. Little America. *Aerial Exploration of the Antarctic, the Flight to the South Pole.* New York, G. P. Putnam's Sons, 1930 (p. 408).

11 Ibid., p. 402.

12 Wally Herbert. *A World of Men.* London, Eyre and Spottiswoode, 1968 (p. 222).

13 A New Zealand party under V. R. McGregor resurveyed our stakes in December 1963. Their measurements yielded a rate of movement over a period of 684 days that was within four percent of ours measured over five days. Since Liv was the slowest-moving of all the glaciers surveyed, we are confident that our results for Scott and Amundsen glaciers were no less accurate.

14 Charles Swithinbank. Motor sledges in the Antarctic. *Polar Record,* Vol. 11, No. 72, 1962, pp. 265-269.

15 Charles Swithinbank. Ice movement of valley glaciers flowing into the Ross Ice Shelf, Antarctica. *Science,* Vol. 141, No. 3580, 1963, pp. 523–524.

16 Charles Swithinbank. To the valley glaciers that feed the Ross Ice Shelf. *Geographical Journal,* Vol. 130, Part 1, 1964, pp. 32–48.

Chapter Eight
Armchair Glaciology

The trail breaker is an indispensable ally of the spiritual values which advance and sustain civilization.
                    Roald Amundsen (1872–1928)

A lot of water flowed under the bridge before I again worked with the US Antarctic Program. Six years of it. During that time, I left the University of Michigan for the University of Cambridge from whence I had come in 1959. The move was not from any kind of dissatisfaction with Michigan; in fact, I was very sorry to leave. The reason was that I had become acutely aware that — collectively — western glaciologists were ignoring a great deal of Russian published material simply because of the language barrier. Loathing grammar books, I wanted to learn Russian the easy way, by spending a year as an exchange scientist in the Antarctic. Several Americans had done it already, and I asked to follow them. However, the State Department in Washington understandably balked at the idea of someone with a British passport being passed off as an American exchange scientist.

I next asked the British, through the Royal Society in London, if they would consider arranging an exchange for me. The outcome depended on the Russians, and on finding an employer prepared to do without me for 18 months while at the same time paying a salary into my bank account. Knowing of my ambitions, Gordon Robin, Director of the Scott Polar Research Institute in Cambridge, with some difficulty secured a post for me on the basis of funding provided by the British Antarctic Survey. He made clear that on returning from my sojourn with the Soviet Antarctic Expedition, I would be expected to develop a glaciological research program at the SPRI. To this I agreed.

I spent a happy 18 months with the Ninth Soviet Antarctic Expedition, returning to Cambridge in May 1965. I could write a book about the experience, but Gilbert Dewart, an American exchange scientist, has done a better job than I could.[1]

Back at the Scott Polar Research Institute, the big news was of the recent development of an ice-penetrating radar. Perhaps here at last was an instrument that could complete our work of 1960–

1962 on the valley glaciers feeding the Ross Ice Shelf. Owing to the loss of Ed Thiel, we had failed to obtain ice depths across the glaciers whose rate of movement we had measured. The gravity measurements had yielded approximate ice depths,[2] which later turned out to be good, but there was no way of proving it at the time.

In February 1962 I had flown from McMurdo to Christchurch with Admiral Tyree and Amory Waite. Waite had served as radio technician on the Second Byrd Expedition at Little America in 1934. He noticed that Very High Frequency radio communication was possible through an iceberg, a concept regarded by some as heretical. At the time, ice was generally believed to be opaque at radio frequencies. In 1958 he had made soundings through glacier ice from a sledge and in 1960 showed that radio echoes, or reflections, could be obtained from an airplane flying over the ice. Part of the signal was reflected from the surface and part from the glacier bed. The time difference between the two reflections gave a measure of the ice thickness. Waite's instrument was not designed for the purpose, so could only get reflections where the ice was less than about 1,250 feet thick. Unfortunately we already knew from seismic soundings that most of the Antarctic ice sheet was much thicker.

Waite was employed by the US Army Signal Corps, which was more interested in developing aircraft radar altimeters that would not penetrate ice. The Scott Polar Research Institute decided to build a pulsed radar designed to measure the greatest possible ice depths. The designer was Stanley Evans and his prototype was mounted on a sledge and tested in the Antarctic.[3] Evans saw no reason why it should not work equally well from aircraft.

Geoffrey Hattersley-Smith, a glaciologist with the Defence Research Board of Canada, secured the use of an Otter aircraft to test the concept over Ellesmere Island in Arctic Canada. Robin and Evans brought along their instrument and installed it in the Otter. The trials took place in April 1966 and were eminently successful, recording ice depths of up to 3,000 feet.[4] A film camera had been modified so that film moved slowly across the face of an oscilloscope display graduated in microseconds. The effect was to trace a cross section of the ice sheet over which they flew. In subsequent developments the technique has variously been known as radio echo sounding, radar sounding, and electromagnetic sounding.

The logical progression was to try the instrument in the Antarctic. David Petrie and I went south with it eight months after the Ellesmere Island trials and fitted it in an Otter of the British Antarctic Survey. Later we transferred it to a smaller aircraft, a Swiss-built Pilatus "Porter." In the course of 75 hours of flying time over the Antarctic Peninsula we obtained depths of up to 2,300 feet over land ice and floating ice shelves.[5]

Seismic sounding, the only satisfactory method of measuring ice depth before this, generally required several hours of work to obtain results at even a single point. Imagine our feelings on finding that from each hour of flight, while admiring the scenery from a comfortable chair, we could carry home an almost unbroken cross section covering 100 miles or more of this icy landscape.

On the way home to England I showed our results to Bert Crary at the National Science Foundation in Washington, DC. He was excited by them, and said that he had discussed with Gordon Robin the possibility of mounting an SPRI radio echo sounder in a USARP aircraft the very next season. Because of the limited range of the light aircraft that we had been using, I was delighted at the prospect of once again working out of McMurdo, this time with a real possibility

US contribution to a joint program involving the use of transport aircraft would add up to many times the British contribution. But SPRI had the best instrument and USARP had the best aircraft, so collaboration was the obvious way forward. At first Crary offered the use of an R4D, but on reflection settled for a C-121J Super Constellation. The C-121's four engines would be safer than two, and its range was more than twice that of the R4D.

Gordon Robin and Bev Smith with "our" vehicles. The radar antennas are strung beneath the tail fins of the C-121J Super Constellation.

of obtaining the ice thickness measurements still needed to complete my earlier work.

Crary's offer was particularly generous in view of the relative costs involved. The

The radio echo antennas had to be strung under the tail of the aircraft between its three vertical stabilizers. So complicated was the installation — never before attempted on an airplane of this size — that NSF asked the Navy to fly the C-121 over to England for discussions with SPRI personnel. The aircrew were not averse to enjoying a free trip to Europe. They arrived at Lakenheath airfield, 20 miles east of Cambridge, on 31 August 1967 and spent two days with us. Their newly-painted aircraft had the name *Phoenix* painted on her nose cone. The Captain, Lieutenant-Commander J. K. (Jake) Morrison, USN, was cooperative and enthusiastic about the planned work.

Ten weeks later we assembled in Christchurch, New Zealand. Eddie Goodale and Phil Smith, old friends from years before, were there to greet us. Goodale was in his last season as NSF liaison officer. Having started with Byrd's first expedition in 1928, he was to retire from Antarctic affairs after a career spanning forty years. He and Phil did everything possible to smooth our path through the administrative maze of the USARP/Navy relations. The Cambridge contingent consisted of Gordon Robin, Stan Evans, Bev Smith, and myself. Smith was a young electronics engineer employed by the British Antarctic Survey. Evans had come just to help fit the instruments to the aircraft and to ensure that everything functioned as it should. We had brought two radio echo sounders to allow for uninterrupted sounding in case of any malfunction.

With a sixty-ton aircraft and its eight-man aircrew placed at our disposal, we felt as people must feel after winning the football pools or some giant sweepstake. It was a luxury of which most of us had never dreamed. Jake Morrison introduced us to the other seven members of his crew: lieutenants Ernest Missildine, Jerry Finch, Steve Riley, and Raymond Binder; chief petty officers Glenn Hunt and Robert Taylor; and Petty Officer Mike McGuire. It was not until we reached McMurdo that we realised that they kept another aircrew on stand-by there; the aircraft could fly around the clock without the

Jake Morrison.

crew members exceeding their duty hours.

The first test flight was on 18 November, over the sea near Christchurch. Perhaps to remind us that we were experimenting with the unknown, one of the three antenna pylons beneath the tail broke off and fell into the sea, and on landing, fatigue cracks were found in the fin supporting it. On the second test flight, another pylon broke and its antenna wire parted. So it was back to the drawing board. On the third test flight almost everything functioned. Although there was no ice around, we obtained good radar reflections from the sea surface. As the radio echo transmitter was operating close to frequencies used for radio and television, and we were not licensed to use them, we were anxious not to be identified as the culprits by inadvertently upsetting television watchers in Christchurch. One member of our party stayed in a hotel room to monitor the television in the hope of forestalling any complaints from the populace. He reported that some lines flashed across the screen but that it would be almost impossible to identify our aircraft as the source of the problem. However, we were as anxious as anyone to limit our possibly antisocial behavior by putting 2,000 miles between us and Christchurch.

This we did on 5 December. After a smooth flight to McMurdo, we landed on the sea-ice runway in sunshine seven hours after leaving New Zealand. There are few runways on Earth with a tendency to break up and drift out to sea at an indeterminate date towards the end of summer. But this one does. There is an annual cycle which begins when the waters of McMurdo Sound freeze over in the fall season. Throughout the ensuing winter the ice is growing thicker. By the time the first aircraft arrive from New Zealand in October, it is more than six feet thick and strong enough to bear the heaviest aircraft ever built. Apart from snow-plowing to keep the surface bare, no preparation is needed. By late-December the runway deteriorates, with slush and puddles of melt water. All aircraft without skis must retreat to New Zealand.

Thus our C-121 had at most about three weeks before it would have to fly out of the Antarctic. Sometime in January or February, without any warning, the sea-ice can break into separate ice floes and drift away to sea. Anything left on the ice runway goes with it. In some summers it never breaks up, and the ice grows thicker and thicker during the second or third winter. Nobody has ever found a way to predict the outcome in any particular year.

McMurdo was agog with the news that at the British Halley Bay Station on the opposite side of Antarctica, two men had inadvertently marched over an ice cliff while man-hauling their camping gear in a whiteout. One was seriously injured and the British had asked NSF if an evacuation flight might be possible. The aviators of VX-6 had contingency plans ready for the unexpected and now they sprang into action.[6] Within hours of getting the call an LC-130 was airborne for Halley Bay with a doctor. Refuelling at the South Pole, the pilots heard that the British had marked out a runway with the remainder of their year's supply of cocoa. The casualty was John Brotherhood, the station doctor at Halley Bay. He had broken several vertebrae and lost all his front teeth. Jim Shirtcliffe, who fell with him but suffered only minor injuries, had maneuvered his colleague into a sleeping bag and pitched a tent over him.

When the aircraft landed at Halley Bay, Brotherhood was gently loaded on a stretcher. Taking off without delay, the crew headed for McMurdo to refuel. From there the flight continued with a relief aircrew to New Zealand. The patient was safely in hospital in Christchurch 26 hours after the squadron had been alerted to the emergency. In that time the aircraft had flown 6,000 miles. It was a wonderful example of international solidarity in coping with the hazards of life in Antarctica. No charge is ever made for such services. The giver knows that, some day, the tables may be turned. At one time or another, almost every national expedition in Antarctica has been involved in helping another in an emergency.

The Halley Bay affair was not the only emergency that day. Deception Island, a dormant volcano on a small island on the South American side of Antarctica, was erupting and throwing black clouds of ash to an altitude of 20,000 feet. There were three research stations on the island with a population totalling 52: Argentine, British, and Chilean. The Chileans were closest to the site of the eruption and fled overland to the British station. Then ash began falling on the British as well. Helicopters from a Chilean ship rescued first the Chileans and then also the British personnel. The Argentines were rescued from the other side of the island by helicopters from an Argentine ship.

We expect to live with hazards in the Antarctic but this one was not on the menu. McMurdo lies in the shadow of Antarctica's only other active volcano, Mount Erebus, leading some of the residents to wonder whether they too might one day be bombarded with ash and volcanic bombs.

One of the great strengths of USARP is that it has always welcomed suitably qualified foreigners — without requiring them to share the costs. I had been the beneficiary of this policy in three earlier seasons, and was now considered almost an honorary American. The fact that I was known to have an American wife was regarded as a bonus. This season there was to be one participant from Australia, three from Belgium, two from Chile, one from West Germany, three from Norway, one from South Africa, the three of us from the United Kingdom, and two from the USSR.

The winter population of McMurdo Station was to be only 76. Now, in November, the total population numbered more than 1,000. The place was a hive of activity, with a "personnel" building of immense proportions under construction. The nuclear power station was celebrating its sixth year of operation. Its staff boasted that it had maintained 86 percent "availability" during the past year. This struck me as a triumph of public relations. Not everyone grasped the meaning that the station had been "unavailable" (i.e., broken down, or refueling) 14 percent of the time. However, the annual fuel saving compared with the old diesel-electric power station was impressive: 1,870 ton. Years later I learned that an 86 percent availability was better than nuclear power stations back home, so it really was an achievement. A new sea water distillation plant using waste heat from the nuclear reactor was producing 110 ton of fresh water daily. Showers were allowed twice a week instead of once.

Six US stations were now in operation: McMurdo, South Pole, Plateau, Palmer, and Hallett (jointly with New Zealand). We learned of the imminent departure from Antarctica of the last of the R4D aircraft that had done so much for me six years earlier. The same type had been in Antarctic service since the Navy's Operation Highjump in 1947.

The NSF representative at McMurdo was Jerry Huffman, a young and keen outdoorsman. He got us settled in while at the same time recognizing that, because of the short season, our only thought was to get flying. We were handed the keys of a half-ton truck to use as our personal runabout for the time we were at McMurdo. Gone were the days of waiting to hitch a ride. We made a trial flight the very next morning. It showed that — at

The upper reaches of an Antarctic Ice Stream. This is David Glacier from an altitude of 15,000 feet. The ice flows from top right to bottom left of the picture.

last — everything was working. Landing in the late evening, when ordinary mortals would have opted for bed, we asked to refuel and get going. Jake Morrison was unperturbed and we were airborne at midnight.

We headed west from McMurdo, over the dry valleys and then on to the rather

With a professional navigator on board, it was a delicate task to explain this without implying that his navigation was not good enough. Pilots and navigators are skilled at guiding an aircraft to its destination, but in the Antarctic at least, they may be less concerned about their exact position at every stage of the flight. We

The two aircrews of *Phoenix*.

featureless ice sheet beyond. The object was to find the deepest ice in order to test the performance of our instruments. We explained to Jake Morrison that it was important for us to record the aircraft's exact position in relation to mountains or any other visible features, and that this could only be done if one of us sat in the cockpit. The view out of the side windows was too restrictive.

needed to know where we were all of the time. When no mountains were in sight, we relied totally on the skill of the navigator wielding his bubble sextant. Jake Morrison understood our needs, and throughout the season, Robin and I alternately occupied the copilot's seat. Any mild objection from the displaced copilot was muted by the presence of a number of bunks for off-duty crew; not to mention a mini-library of paperbacks.

Bev Smith, meanwhile, was crouched over a vast array of instruments that we had brought from Cambridge. An impressionable young man, he appeared almost shell-shocked by his sudden immersion in this alien world. Robin and I, in contrast, knew what to expect. We had worked together in the Antarctic 15 years earlier (1949–1952) and had been collaborating in one way or another ever since. This was my eighth summer season in the Antarctic and fourth working out of McMurdo, and many of the staff were friends.

Bev's strength lay in his ability to analyze and cure instrument malfunctions within minutes. Three film cameras had to be kept running continuously, so that while one was being reloaded there were no gaps in the record. A mass of cables snaked between separate units on giant racks. A 20-channel recorder kept a continuous trace of the aircraft's heading, height, and airspeed; the outside air temperature; and many other things that we needed to know. You could tell that members of the aircrew looking at the array were restraining themselves from exclaiming "What a mess!" We explained that while tidy instrument lay-outs may be good for marketing, ours was intended simply to be functional and was not for sale.

Besides Bev Smith, the only other occupant of the vast passenger cabin was a 2,000-gallon rubber fuel bladder. Some people called it a pillow tank because of its shape. The extra fuel allowed us anything up to twelve hours in the air while still maintaining a healthy reserve.

One thing did give us pause. With wheels but no skis on the aircraft, McMurdo was the only place to land. Come what may, we had to find our way back there. I remembered the uncomfortable moments in late October 1960 when, as a passenger in this same aircraft, we had arrived at McMurdo from Christchurch with visibility on the ground down to 100 yards. I asked Morrison what he would do if we returned to find the airstrip in fog. He replied that in that case his orders were to ditch (if that is the word) wheels-up somewhere on the Ross Ice Shelf. Being level and well clear of mountains, the aircraft might be a write-off, but in theory at least, the crew would simply walk off, and hopefully get rescued by a ski aircraft before they froze to death.

I remembered the quite extensive snow-free blue ice areas that we had found in 1961–1962 on Scott Glacier, and I had seen others on the inland side of the dry valleys. Some of them were smooth enough and long enough to allow safe landings and take-offs in the C-121, so I explained this to Morrison. But orders, I was reminded, were orders, even in VX-6.

Sitting up front, Robin and I were in the best position to talk to the pilot. There soon developed a relaxed and cheerful atmosphere; our little group was allowed to feel very much a part of the aircrew. We never had any problem communicating our needs, and there was no difficulty in understanding the crew's concern with their engines, the fuel supply, and the weather at McMurdo.

After the trial flight, we landed back at McMurdo at 0500. That we already aspired to the most accurate records is shown by Robin's note in the flight log:

Check clock times on this flight — possible 2 second adjustment.

Flying, as we had been, at 230 knots, we were covering 130 yards every second — so seconds counted.

While the crew went off for a well-earned rest, Robin and I hurried along to the Navy photo laboratory. The photographers had agreed to develop our film records immediately, so that any problem could be identified and we could deal with it at once. Throughout the season they kept a photographer in the lab, day and night, on stand-by.

There were many other watch-keepers. Meteorologists; flight controllers; fuelling teams; airframe, engine, and avionics engineers — all of them were involved in keeping us flying safely and all of them had to be available all of the time. We felt humbled by such bountiful attentiveness, all of it planned to allow us to get on with the flying.

After 10 hours on the ground, we were off again. Flying east to the 180th meridian, we then turned south, traveling down the middle of the Ross Ice Shelf to its southernmost extremity at the foot of Reedy Glacier. Taking over the log from Robin at the half way mark, I was embarrassed to become "temporarily misplaced" — every pilot's euphemism for getting lost. Some mountains looked familiar but I was disoriented. However, we had turned for home along the Transantarctic Mountains and soon reached Scott Glacier, where Tom Taylor, Art Rundle, and I had traveled in 1961. Looking down on the scene from this new perspective, I could not help feeling like some Godlike creature who had at last deciphered the mystery of flight. The contrast was such that now, speeding along at 200 knots in a comfortable chair, I was but a disembodied explorer. It was all so easy, and infinitely more serene than our earlier incarnation — trudging over the ice at four knots.

There was no time for dreaming. Scott was the first of the glaciers that I had earmarked for sounding. Like a bomb aimer, I guided Jake Morrison to fly precisely over the line of stakes that we had set out six years before to measure the rate of movement. I wondered whether we had found over-cautious pilots who would not fly low enough to avoid any confusing radar reflections from the mountains instead of from the glacier bed. We needed to fly below 1,500 feet. I need not have worried. Accustomed as he was to long hours of boredom at altitude, Morrison was enjoying it, and so was I.

It was anything but relaxing for our colleagues in the back, with only side windows to look out from. At the end of each sounding run, not knowing what to expect, they heard the sudden surge of engine power, felt the machine roll into a steep bank, and watched in horror as ice and rocks flashed by just off the wingtip. It was not what they expected of a four-engined airliner.

Continuing along the Queen Maud Mountains, we ran cross sections over Amundsen Glacier and then Liv Glacier. I could spot each one of the survey stations where, years before, we had laboriously built rock cairns. By the time we reached Beardmore Glacier, Morrison was becoming quietly exasperated at the frequency of my directions, which must have sounded too much like peremptory commands. He threw his hands in the air, saying "If you think you can do better . . ."

Whether or not he knew that I was an amateur pilot, I shall never know. But it was a challenge not to be passed up. I took the controls while he drank a cup of coffee that had just been handed to him. I said: "Take as long as you like over your coffee." A twinkle in his eye showed that he knew what I meant.

My first 180-degree-turn onto the Beardmore stake line unintentionally lost 500 feet of altitude; Morrison pretended not to notice. It transpired that the flight engineer was alternately extending the engine cooling flaps on opposite sides of the aircraft to upset the trim — to my consternation and the aircrew's innocent merriment. It was typical of the jocular atmosphere in the cockpit. Fortunately, my clumsy flying turned out to be a blessing in disguise. Now down to 1,000 feet above the glacier, we got the best cross-sections of the whole flight.

An hour later there were more good soundings across Nimrod Glacier. When we finally landed at McMurdo after midnight, we had been in the air 9½ hours. We spent the following two days looking over the results recorded on film.

The next flight, on 11 December, was to the mighty Byrd Glacier, the biggest in the Transantarctic Mountains. Cross sections involved flying 1,000 feet over the ice straight towards Horney Bluff, our survey station of 1960–1961 sitting at the top of a cliff rising to 3,700 feet. Wishing our sounding line to run the whole width of the glacier, I asked Morrison to head straight for the cliff. Faced with a massive black wall of rock ahead, my heart skipped a beat as we closed in at 200 knots. Just as I was beginning to feel that he was bent on suicide, he asked "What if the controls failed now?" In such moments it is good to know that the pilot, at least, has kept his sense of humor. At this point the flight engineer appeared in the cockpit, perhaps to reassure himself that we were not dozing. Morrison barked: "METO power!" (Maximum Except for Take-Off) and threw the aircraft into a steep bank. Again I felt for those back aft, who no doubt wondered whether the rapid surge in power and tight turn heralded some emergency.

There was no doubt that Morrison was having fun. I began to understand why airline pilots, condemned to a lifetime of sedate maneuvering with loads of anxious passengers, are often seen at small airports on days off throwing their private aircraft around the sky. They had taken up flying for the love of it, only to discover that earning a living by it brought hour upon hour of autopiloted tedium.

Jake Morrison's autopilot had failed some time ago. He demonstrated the problem by switching it on for a moment, at which point the control column was wrenched from his hands and the aircraft nosed into a dive. Manual flying felt safer.

Leaving Byrd Glacier, we flew two cross sections of Mulock Glacier and one over Skelton Glacier on the way home. Now the crew were in the swing of it. On seeing the results of the flight, Robin decided that 1,000 feet over the surface was the best height to fly, except over ice shelves which were quite easy to handle from 4,000 feet.

The next three days saw us in the air for 29 hours, yielding 5,000 miles of new data. Robin and I were ecstatic, remembering that when we were seismic sounding 15 years before, it had felt good to discover the depth of ice at a single point at the end of a hard day's work.

Taylor Valley from an altitude of 19,000 feet, with McMurdo Sound and Ross Island in the distance. Taylor Glacier (foreground) flows toward the ice-covered Lake Bonney.

On 15 December we went into the dry valleys on the west side of McMurdo Sound. The geologists working there must have been startled to see a four-engined airliner cruising up Victoria Valley well below the hilltops. Turning right through Bull Pass, we ran into severe air turbulence. Enquiring over the intercom how Bev Smith was faring I got the reply, "We are hanging on for dear life!"

We continued by flying an ice thickness profile along Erebus Glacier Tongue at the request of Gerald H. (Jerry) Holdsworth, a New Zealander who had been studying it for some years. Most expeditions in the area since Scott's 1901–

1904 *Discovery* expedition had mapped this floating tongue but nobody knew how thick it was. The tongue advances slowly and then calves suddenly, and has varied in length over the years from six to 10 miles. We found it was 1,000 feet thick where it flowed off the land but less than 300 feet thick at the far end.

Next, at the request of the zoologist William (Bill) Sladen, we made a couple of photographic runs over the Adélie penguin colony at Cape Crozier. Using high resolution large-format photographs from the aircraft's vertical survey camera, Sladen proposed to do a periodic census of all penguin colonies in the vicinity. Finally, on returning across the open water of McMurdo Sound, Robin asked to fly low to calibrate the sounder. One recording device later showed how accommodating our pilots had been: they had descended, albeit briefly, to a height of 37 feet over the sea. The tips of the propellers must have been lower.

On 19 December we prepared for a long flight over the higher parts of the Antarctic Plateau to the west of the mountains. That meant loading 6,550 gallons of fuel in the main tanks and an extra 2,000 gallons in the passenger cabin. Heading for Vostok, a Soviet station 11,500 feet above sea level, we kept 1,000 feet over the surface to get the best radio echo results. Vostok is the spot where the world's lowest temperature, -129° F, has been recorded.

Sitting in the copilot's seat was no rest cure. Our job was to back up the automatic recording devices by noting the time, to the nearest second if possible, of all events such as heading changes and positions. At least every five minutes throughout the flight we logged the airspeed, altitude, outside air temperature, and height above the ice surface. Also bearings to mountains, the alignment of snow surface sastrugi, the presence of crevassed areas or other surface topography.

Morrison was intent on buzzing the Russians by flying over them at 200 feet. Having spent a day at Vostok four years before, I warned him that there were some very tall radio masts. In return I got the look that back-seat drivers get when they inform the driver that there is a STOP sign ahead.

The Russians, living on their frigid plateau and more isolated than any other community on Earth, were probably glad to have something new to look at for a minute or two as we circled. I called them repeatedly on the radio in my best Russian but there was no response. Their runway was conspicuous by its length — 2½ miles — a good indication of how difficult it is to get airborne in the rarefied air.

An hour after leaving Vostok we homed in on a tiny drifted-over hut, the site of the former Sovetskaya Station. On one end of the hut there was a bust of Lenin heroically defending this icy wilderness from infidels. Beyond Sovetskaya we headed towards the so-called Pole of Relative Inaccessibility — the area farthest from any point on the coast. At one stage we had to climb to almost 15,000 feet to keep clear of the surface. The outside air temperature was -18° F.

Thinking of contingencies, as pilots are wont to do, Morrison casually asked the flight engineer to refer to the flight manual to see what our "three-engined ceiling" would be at the aircraft's present

USS *Atka* after a blizzard at McMurdo.

weight. It was disconcerting to be told that it was 13,000 feet. This meant that in the event of failure of any one of our four ageing piston engines, we would inexorably descend to 13,000 feet. The snow below us being at that altitude, we would — in plain words — crash.

What I did not know at this time was that, ever since passing over Vostok, the radio operator had failed to make any contact with McMurdo. Morrison knew, and he also knew that according to the squadron's standing orders, he should have turned for home immediately. Infected by our enthusiasm for flying over the Gamburtsev Mountains, a jagged subglacial range in places reaching almost up to the ice surface, he had continued to fly further away from McMurdo. When we did finally turn for home, McMurdo had not heard from us for three hours.

We found out later that the flight controllers at McMurdo, knowing that our flight plan would take us over the highest part of the ice sheet, had also looked at their flight manuals to find our three-engined ceiling. Taking this and the radio silence, they understandably feared the worst. Well-rehearsed emergency procedures sprang into action. An LC-130 on a fuel run to the South Pole was diverted to search for us. In order to reach Vostok, the crew had to pump their cargo of fuel into the aircraft's own wing tanks. They too could make no radio contact with the Russians, so landed at Vostok to check whether the Russians had any ideas about where to search. They only returned when McMurdo recalled them. Another LC-130 took off from McMurdo to search a different area.

Unable to transmit, we could hear all this activity over the radio. The more we heard, the more our embarrassment grew. We were still hours away from McMurdo and there was no way to tell them that we were homeward bound and that everything — except the radio — was functioning like clockwork. Our short-range (VHF) radio was still in good order, so the crew kept up a stream of calls to anyone who might be within range. Eventually one of the search aircraft heard us. Their crew sounded immensely relieved to find that we were still in the air and that we knew where we were. We gave them our expected time of arrival at McMurdo and they intercepted us as we reached the ice shelf. With an escort off our wingtip, we landed 12 hours after starting out.

It took only a glance to realize that the ground crew were well aware that we had broken the rules. There was to be no mercy. Everyone was tight-lipped and there was a helicopter waiting to take Morrison to face the Admiral.

Morrison had transgressed, but he had done it all for us. As far as the Navy was concerned, the aircraft's commander was solely responsible. There was nothing we could do to take the heat off him. Later, we told the Admiral what exciting results we had obtained on the flight, obliquely hinting that we had all been carried away. Though we could not say it, he sensed what we were driving at. To us, Morrison was more hero than sinner. As far as I know, no lasting damage was done to his career.

After two days spent analyzing results, we were off again on 22 December. This was a triangular flight to the northwest, climbing to 10,000 feet over the plateau and reaching the coast at Ninnis Glacier, due south of Tasmania. Christmas Eve saw us flying in the opposite direction, to Byrd Station. The interest here was that glaciologists were about to complete a drill hole right through the ice sheet and expected to strike bottom at a depth of 7,100 feet. The opportunity to compare depth measurements was not to be missed. Our method depended on knowing the velocity of radio waves in ice, which we had calculated but only on the basis of unprovable assumptions. Their method was direct and incontrovertible. Unfortunately, at the moment of passing over the drill site the radar bottom reflection faded. However, we were comforted to note from ice depths nearby that the two measurements were close.

Christmas was an enforced rest day because understandably, the aircrew were counting on a day off. Robin and I were constantly aware that our short, frenzied, exhausting, but highly productive flying season was drawing to a close. The sea-ice runway was glistening with meltwater and it was time for the last of the wheeled aircraft to wend its way homeward. Ours was the last.

Christmas day brought church services for all who managed to squeeze into McMurdo's tiny non-denominational "Chapel of the Snows." Afterwards, there was an orgy of good eating. The cooks in the "galley" had excelled themselves. There was everything, it seemed, that anyone's mother had ever put before her child. The valve on the ice cream machine jammed open at one point, causing consternation on the face of the man who was holding his bowl beneath it. Enough ice cream for a dozen people spilled onto the floor before the machine was stopped. I have known people surprised to learn that anyone eats ice cream in the Antarctic. But if they knew how hot the McMurdo mess hall was kept, they would understand that cold showers and ice cream were not always enough to cool the blood.

Someone roused our faithful aircrew from their slumbers the next morning. Evidently they were as keen as mustard to give us a memorable last flight. We were airborne at 0900 and flew for 12 hours. The flight took us over the unexplored eastern boundary of the Ross Ice Shelf where several huge ice streams flow in from Marie Byrd Land. From there we flew to a point only 150 miles from the South Pole before turning for home and descending over Nimrod Glacier. It was a fitting finale. We were all very tired but also exhilarated at successfully measuring ice depths along thousands of miles of flight tracks.

We had spent a total of 94 hours in the air in the space of three weeks. The film records yielded an almost continuous cross section of the ice sheet where ice depths were less than 6,500 feet and intermittent reflections over thicker ice. The colder the ice, the deeper the radar penetrated. The greatest depth measured — nearly 14,000 feet — was over the coldest part of the high plateau. An unusually strong bottom reflection at this depth suggested that there might be water beneath the ice — in other words a sub-glacial lake. The idea of fresh-water lakes beneath the coldest ice in the world was astonishing. Soviet glaciologists had reported that the top layers of snow in the same area indicated a mean annual temperature of around -100° F, yet now we were claiming that the bottom of the ice could be 127° F warmer. As it turned out, both claims have stood up to scrutiny. Proof will be hard to come by owing to the enormous cost of drilling through great depths of ice. Another exciting discovery was of continuous sedimentary layering through more than half the total ice thickness. Well-defined layers could be seen to be deformed upward as the ice flowed over mountain ranges beneath the ice.[7]

Bert Crary at NSF was so pleased that he at once began planning for further seasons of collaboration with the Scott Polar Research Institute.

Analyzing the results proved to be a daunting task. None of us could have done it alone, not even in a lifetime of work. Robin's policy was to recruit an army of Ph. D. students to work on particular aspects. All of them had reason to be grateful for the long hours that we had spent in the air. A raft of publications later described the findings. We had been privileged witnesses at the birth of a new science — *Radioglaciology*.[8]

---

[1] Gilbert Dewart. *Antarctic Comrades.* Columbus, The Ohio State University Press, 1989.

[2] M. Giovinetto, Edwin S. Robinson, and C. W. M. Swithinbank. The regime of the western part of the Ross Ice Shelf drainage system. *Journal of Glaciology,* Vol. 6, No. 43, 1966, pp. 55–68.

[3] J. T. Bailey and S. Evans. Radio echo-sounding on the Brunt Ice Shelf and in Coats Land, 1965. *British Antarctic Survey Bulletin* No. 17, 1968, pp. 1–12.

[4] S. Evans and G de Q. Robin. Glacier depth sounding from the air. *Nature* (London), Vol. 210, No. 5039, 1966, pp. 883–885.

[5] Charles Swithinbank. Radio echo sounding of Antarctic glaciers from light aircraft. *International Association of Scientific Hydrology*, Publication No. 79, 1968, pp. 405–414.

6 Evacuation flight to Halley Bay. *Antarctic Journal of the United States*, Vol. 3, No. 1, 1968, pp. 14-15.

7 G. de Q. Robin, C. W. M. Swithinbank, and B. M. E. Smith. Radio echo exploration of the Antarctic ice sheet. *International Association of Scientific Hydrology*, Publication No. 86, 1970, pp. 97-115.

8 V. V. Bogorodsky, C. R. Bentley, and P. E. Gudmandsen. *Radioglaciology*. Boston, D. Reidel Publishing Company, 1985.

Chapter Nine

Helicopters Unlimited

Women will not be allowed in the Antarctic until we can provide one woman for every man.[1]
Rear-Admiral George Dufek, USN, 1957

Eleven years elapsed before I was once again at McMurdo. During that time I spent three summers in the Antarctic with the British Antarctic Survey, steamed through the Northwest Passage in SS *Manhattan* — the largest ship ever to venture into pack ice — and voyaged to the North Pole in HMS *Dreadnought,* a nuclear-powered submarine. But most of the time was spent in Cambridge. The British Antarctic Survey appointed me Head of Earth Sciences, with the task of building up a team of geologists, glaciologists, and geophysicists to tackle ever more ambitious research projects in the Antarctic. This meant drafting proposals that could pass scrutiny by committees of learned scientists, then supervising everything from training, through field work, to the preparation of results for publication. I felt happy and productive — though never so happy as when taking part in field work.

In August 1978 an invitation came out of the blue from Terence J. (Terry) Hughes, an eccentric professor in the Department of Geological Sciences at the University of Maine in Orono. Terry planned to visit one of my old hunting grounds — Byrd Glacier — to enlarge on the work that I had done 18 years earlier. He thought my experience could be useful. He had assembled a group of young scientists that included three geologists, a physicist, and an engineer. Terry himself was originally a metallurgist but now was considered more of a physicist. The object, as in 1960, was to measure ice flow velocities. But this time he proposed to measure the velocity at hundreds of points spread over the whole length of the glacier, an area of 1,500 square miles. His interest lay in trying to understand the dynamics of fast glacier flow. There was no better place than Byrd, the fastest-flowing glacier in the 2,000-mile-long Transantarctic Mountains.

Fast glacier flow is believed to have played a critical part in the dynamics, stability, and later disappearance of ice-age ice sheets in the Northern Hemisphere. Hopefully, we could test the hypothesis by learning what factors controlled the stability (or instability) of an Antarctic ice stream/outlet glacier/ice shelf system. It was evident that Byrd Glacier straddled the grounding line, the area where the ice flows off the land and on to a frictionless medium — the sea. Could the grounding line be shifting with time, and if so, what would be the consequences? If the climate is now changing, how much can it change without destabilizing glaciers like Byrd?

Hughes' team consisted of Henry Brecher, an experienced topographic surveyor and, at 46, the oldest man in the party apart from myself. Terry was 40 and had worked in the Antarctic before. James (Jim) Fastook was a 29-year-old physicist whose mind worked best with abstruse mathematics. Tad Pfeffer was a 26-year-old geologist, a mountain of a man whose passion was mountaineering. Jeffrey Lingham (on loan from another project) and Mark Hyland, both geologists from the University of Maine, were the youngest members of the party.

From the US Geological Survey we had arranged to borrow Thomas (Tom) Smith and Leland (Lee) Whitmill, both experienced surveyors in their thirties. Survey was important because we needed a network of ground control points to get the best out of aerial mapping photographs that were to be obtained during the season. The only existing map of the area was described as a "reconnaissance" map. As preliminary maps go, it was very good indeed, but we needed something more precise.

Terry Hughes was a man of ideas in an interdisciplinary science. He had a knack of throwing caution to the winds in advancing outrageous hypotheses without a shred of evidence to support them. But in doing so he aroused his colleagues to think beyond the narrow confines and unwitting prejudices of their fields. He could apply an unusually broad and thorough grounding in physics, mechanics, and mathematics to new problems. Yet he was intellectually — and in almost every other way — undisciplined. I was not to realize how undisciplined until we spent a month together in a two-man tent.

The other side of his character was that he was generous, to a fault, in his dealings with people he respected — and thus vulnerable if anyone let him down. He did not suffer fools gladly, and his transparent contempt for some of the military did nothing to help our cause. What was good was his unbridled enthusiasm, and those who shared in it were carried along.

His sense of humor was blended with bravado. As a construction worker in South Dakota at the age of 15 he had agreed, for a bet, to be tumbled inside a cement mixer with a load of cement, sand, and water. A truck driver was to pay him a penny for every revolution. The only thing that saved Terry from serious injury was the cushioning effect of the cement and sand as he was thrown against the blades inside.

Terry was rotund. He kept in shape — his shape — by satisfying a voracious appetite. On earlier visits to McMurdo he had not been satisfied with the three main meals, so he topped up with the fourth main meal — served for the benefit of night shift workers — at midnight. One day a helicopter, with Terry as passenger, failed to get off the ground. The pilot asked Terry to get off, after which the machine took off without difficulty. From then on he overheard pilots warning of "max cargo" whenever he approached a helicopter.

I met up with some of the team on 2 November at the Naval Air Station at Point Mugu, California. With us was Alfred (Al) Fowler, an ex-Navy Captain who, like others before him, had become so enamored with Antarctica that, on retiring from the service, he went to work for the National Science Foundation. Now he was deputy Director of the Division of Polar Programs. The presence of two other foreigners, Bjørn Andersen from Norway and Henry Rüfli from Switzerland, showed that the US Antarctic Program had lost none of its attitude of accepting foreign participants.

We were herded into a US Air Force Lockheed C-141, a giant transport with ear-splitting jet engines and no windows. It flew to New Zealand with refuelling stops at Honolulu, Hawaii, and Pago Pago, American Samoa. Welcomed to Christchurch by Walter (Walt) Seelig, who had long since replaced Eddie Goodale as NSF liaison officer, we were driven to hotels to rest. Walt was formerly at the US Geological Survey and knew a lot about maps. Like Goodale before him, he did everything possible to make us feel welcome. The whole operation had expanded since my last visit. We were handed a nine-page booklet, *New Zealand Arrival, Processing, and Departure Instructions,* and a map of the "US Antarctic Program Complex," a two-acre group of administrative and accommodation buildings next to the airport.

Two days later, after collecting cold weather clothing, it was off to McMurdo in another C-141, 60 of us facing each other on opposite sides of the cavernous interior. Instead of seats there were canvas slings supported by aluminum tubes; sleep was impossible except flat on the floor, which was too cold. The only thing one could be grateful for was that the ordeal was over in six hours.

On landing at the sea-ice runway we were loaded into a bus to be driven to the McMurdo mess hall for food. It was in the vast "personnel" building that I had seen under construction 11 years before. The official name was McMurdo Station Dining Facility. The menu, with a penguin cartoon on the front, trumpeted "Every Day a Holiday — Every Meal a Banquet!" By my humble standards, the boast was justified.

After food, the newcomers were assembled for briefing in the NSF "Chalet," the only building in the whole settlement for which architectural aesthetics had played any part in design. It was a passable reproduction of the Swiss style. There was no shortage of information at the briefing. We had already been given a 72-page *US Antarctic Program Personnel Manual* which told everything we needed to know and a good deal more. In advising on Wills, it counseled: "You are urged to have one prepared before you depart for Antarctica." But with *Survival in Antarctica* (66 pages), death was to be avoided at all costs. Now we were handed *Your Stay at McMurdo Station Antarctica* (16 pages) and *Helicopter Operations and Safety Guide* (10 pages). Gone were the days when anyone could plead ignorance. Now the system seemed geared to a generation with a suburban background and no wilderness experience, or to people who might sue the government if, after an accident, it could be shown that they had not been told of some material fact. An instinct for survival was no longer enough.

Accommodation was in double rooms at the Mammoth Mountain Inn, with central heating, sheets on sprung beds, hot and cold running water, and flushing

The nuclear power station that was functioning on my last visit in 1967 was no longer to be seen. It had been packed up and shipped home. Evidently, the fact that the back-up diesel power station could not be dispensed with made the economics of maintaining a nuclear reactor debatable. Now there was a great

The NSF "chalet" overlooking a bust of Admiral Byrd and the flags of some Antarctic Treaty states.

toilets. This was a far cry from the bumfreezer. Next door was the USARP Hotel, already full of summer visitors like ourselves. At the Berg Field Center, named after a geologist killed in a helicopter crash, we were equipped with camping gear. Terry Hughes made up some food boxes. The contents were not even a distant cousin to field rations; more akin to a field surfeit. Gluttony was to have free rein.

hole in the ground where the reactor had been. It had taken six years, from 1972 to 1978, to complete the removal of radioactive layers of rock so that no trace of contamination remained.

It was early in the season, but already the population of McMurdo was about 800. What was new was that a number of them were women. This was the beginning of a striving for equality of the sexes which took some years to take hold.

Antarctica had always been regarded as a man's province. Indeed I met some who came south to escape from women. Wally Herbert had published a book under the title *A World of Men*.[2]

Two members of the Ronne Antarctic Research Expedition of 1947–1948 had been women. Finn Ronne's wife Edith, and Harry Darlington's wife Jenny, spent the winter of 1947 at Stonington Island.[3] However, that was a private expedition. Governments are more conservative. The long uphill battle has been well documented by Elizabeth Chipman.[4] Her book *Women on the Ice* is, among some Antarcticans, unchivalrously referred to as *"Frigid Women."* Admiral Byrd was quoted as saying that the reason Little America was the quietest place on Earth was because no women had ever set foot there.[5] Admiral George Dufek, who commanded Operation Deep Freeze in 1957, was strongly opposed to the introduction of women.

But two years later he was fighting a rearguard action. Conceding that women might soon be allowed to visit American bases in the Antarctic, he was reported as saying: "I felt the men themselves didn't want women there. It was a pioneering job. I think the presence of women would wreck the illusion of the frontiersman — the illusion of being a hero."[6] The French Antarctic leader Paul-Émile Victor[7] was quoted as saying: "We already have enough worries and I see no reason why we should help to create new ones."[8] As late as 1965, Admiral Reedy was reported to have said that, as far as he was concerned, "Antarctica would remain the womanless white continent of peace."[9]

Yet history records that men's attitudes have not always been so negative. Indeed quite the reverse. Robert Edwin Peary (1856–1920), who spent his life trying to reach the North Pole, wrote: "Feminine companionship not only causes greater contentment, but as a matter of both mental and physical health and the retention of the top notch of manhood is a necessity."[10] Peary underscored his point by taking an eskimo "wife" in Greenland, though he was already married, and fathering her child.

It was not until October 1969 that the first woman was allowed to work at McMurdo and in the field. One Antarctic veteran commented: "The only place left now is the Moon!"[11] In 1974 the first woman wintered at McMurdo; she was made Chief Scientist of the wintering party.

It took some years for the old guard to adjust. Here is a sanitized version of a spoof letter addressed to Bert Crary, now retired but still serving as a consultant to NSF:

When God created this Earth, He carefully put in something for everyone; the sea for those who like waves, and hills for those who like to climb. There are plains for those who like to ride horses, farmlands for those who like to work hard and enjoy the masculinity of getting dirty. He put rivers so that there could be big cities — as some people are lost without company. Then He spun the Earth, tilted its axis, and created different climates, so there was happiness for the Eskimos, happiness for the Arabs. But all the time He was doing it, He said, "This little bit down here will be preserved for ever for men. We will make it so miserable that there will be no gift shops nor hairdressers for the women, and it will remain Valhalla forever." To make sure that men came, He put in some fossils here, a coal seam there, made temperature inversions and wind spirals, and tossed in an auroral ring for a climax. This was the Almighty's single greatest effort, much more profound than when He parted the Red Sea.

Now, Albert Paddock Crary, you have goofed it all up.

But now too, the climate was changing. There were women at McMurdo working for the National Science Foundation, women Navy personnel, and women staff of Holmes and Narver Inc., the contractor providing support services. And for the first time, a woman was to winter at the South Pole.

The real face of equality was something of an anti-climax. Women were rapidly accepted as equals. They dressed like men, worked like men, and had no special privileges. A few swore like men, but mostly their presence served to clean up the men's language. If there were sexual liaisons, they were inconspicuous and of little interest to the rest of the population.

Terry Hughes' party, all male, was ready to leave McMurdo for the field on 11 November. Fearing that learned academics — as we were thought to be — might blunder into crevasses, the authorities had given us two New Zealand field assistants, Bill King and Peter Radcliffe. Bill worked for the Post Office at New Plymouth and Peter was a computer manager. Both were experienced mountaineers. Peter, aged 32, had never been to the Antarctic before, but among many other adventures, had climbed in Patagonia with Eric Shipton.

Together with all our supplies, we were loaded into an LC-130 Hercules to be flown south. These big aircraft were now owned by the National Science Foundation but operated, on their behalf, by the Navy. The deluge of information and instructions continued: we were handed a six-page pamphlet *Welcome Aboard the LC-130 Hercules.* Among other things, it told us that aircraft of Antarctic Development Squadron (VX-6) had first landed at McMurdo twenty four years earlier, in 1955; and that now the squadron, "The World's Southernmost Airline," had been renamed VXE-6. The pamphlet also told us about such cheerful things as propeller danger areas, the "turbine disintegration zone," crash-landing, ditching, fire, depressurization, and bailing out. I wondered how many people would travel on civil airliners if they were reminded that these things can happen on any aircraft. But here in the Antarctic it was right and proper that we should be told. Like Boy Scouts, we must be prepared.

Landing on Byrd Glacier was out of the question owing to its chaotic surface, so a base camp had been established on a smooth and level step on Darwin Glacier, 30 miles to the north. We landed in a whiteout to find five Jamesway huts, a bulldozer, and a heavy forklift. Three twin-engined Bell UH-1N helicopters were neatly parked in a row. Camp was not the word that sprang to my mind in surveying the scene. There was accommodation for up to 58 people, including the three helicopter crews, their ground staff, and cooks. Besides the mess hall, there was a radio room, washroom with showers, and flush toilets. Clothes-washing machines and a games room completed the picture. John Splettstoesser, a geologist with years of experience of Antarctic field work, was camp manager.

We were not the only "science project" raring to go. A separate group from the University of Maine at Orono, this one headed by George Denton, was looking for geological evidence of fluctuations in the thickness of all glaciers within helicopter range of the camp. Although the helicopters were capable of operating at twice the distance, the squadron commander decreed that the radius of operations was to be kept within 100 nautical miles. Other science groups were doing isotope and geochemical studies, mapping the structure and stratigraphy of the basement rocks, and studying the intrusive rocks that are widespread in the area. In all, there were about 50 scientists or field assistants involved with 12 separate projects.

Each of these projects had been allocated a certain number of helicopter flying hours, agreed on the basis of plans and discussions that took place before the field season. We were taken aback to discover that 40 hours of our own helicopter time had already been consumed in training and reconnaissance flights — without Terry's knowledge and without any obvious benefit to science. They had scheduled another reconnaissance flight, this time for us, despite my protestations that I had no need to see Byrd Glacier before starting work.

These were not small helicopters. With two pilots and up to six passengers, they weighed 4½ ton each. On a normal flight they carried just over a ton of fuel and 1,600 pounds of useful load. They cruised at 100 knots but could go faster with a light load. Lieutenant Charles Gausseren, USN, the chief pilot, gave us an alarming description of the terrors of flying over the glacier, let alone landing on it. He took off with Terry and Henry Brecher, while I was given a second helicopter with George Denton. In contrast to their boss, our pilots said that there were a thousand places to land. The truth, I knew, lay somewhere between.

It was an overcast day with poor surface definition. However, expanses of bare ice on the glacier, crevasses, and snow patches provided some contrast. We were able to make four rather hair-raising landings some miles from Gausseren's machine. The intercom sounded quite like the first Moon landing, with the third member of the aircrew — the Crew Chief — lying on the floor and leaning out of the door. His advice to the pilots went:

Steady . . . drifting left . . . slip right . . . down two, hold it! Back up . . . steady, slip to your right . . . tail clear . . . down . . . right skid touching . . . left skid on crevasse . . . pull her up, pull up, sir! Okay . . . try again.

The two pilots shared the four landings, two each. Each time they kept most of the weight on the rotors.

The next day, Hughes and I loaded camp gear and 28 man-days of food on one of the helicopters. The copilot was flying under an instructor. It was soon clear that he was a novice — and I was nervous. At one point he lost lift on a narrow spur of ice at the brink of a 3,000-foot precipice. The instructor — and the passengers — were happy when the lesson ended.

Having reconnoitered two camp sites, both with magnificent views, we landed at my old camp site of 1960 at Horney Bluff, 4,700 feet above sea level. Nothing had changed. Three empty fuel drums, two wooden boxes, and an oily rag were just as I had left them 18 years ago at the edge of the rock. After unloading our gear, the helicopter sped off to Darwin Glacier. We pitched the tent, crawled inside, melted snow, and feasted on chili con carne and canned pears.

My most abiding memory of that first evening was the silence. No diesel generator, no aircraft noise, just the occasional gentle rustle of a breeze on the tent. Nobody who has not lived in the Arctic or Antarctic knows what real silence is. So accustomed do we become to noise back home that some people are disturbed by absolute silence. Agoraphobics must never come to this place. To me, silence is one of the most cherished memories of the Antarctic. It is as if we were in on the beginning of the world. Looking at a dazzling panorama of the Britannia Range now spread before us, it was easy to imagine that the scene had not changed since the world began. I felt overwhelmed by the privilege of being alive and being there.

Terry diluted the silence. Mentally operating always in high gear, a constant stream of ideas and stories flowed forth. I nodded and grunted; that seemed to be what was needed. I feel proud that over a month as his constant companion I never said "Shut up!" Probably this was because Terry dispensed a great deal of folk wisdom and I was learning from it.

The night was cold, -10° F. I had trusted others to pack my sleeping bag and that was a mistake. It was the coldest night I ever spent in a tent, though I have camped in temperatures of -40° F. Peter Radcliffe and Bill King arrived in the next day's first helicopter and set up camp beside us.

We had sent a radio message asking for more sleeping bags, and they came with our colleagues. The team was distributed in two camps, one upstream and one downstream from ours. Each of the three camps had a commanding view over the glacier 3,000 feet below. Terry and Peter spent the day inflating bright orange fish floats, each two feet in diameter. It was Terry's novel idea for conspicuous survey targets to be distributed over the glacier.

Each camp had a walkie-talkie radio but the distances were such that, even with binoculars, we could not see the other camps. "Upper Camp" was 12 miles up-glacier, "Lower Camp" eight miles down-glacier. We used signal mirrors — heliographs — to convince ourselves that the camps were intervisible. As long as the sun is shining, the flash of a small hand-held mirror can be seen over at least 20 miles.

Four helicopter flights had been needed to set up the camps. That evening we feasted on filet steak and vegetables. The problem with a gourmet diet in a tent is that it leaves a mess: empty cans, frying fat spattered over sleeping bags, dishes to be washed. Instead of the model system that my University of Michigan group had developed years before, now we needed scouring pads, dish cloths, and a lot more water. The volume of goods needed to prepare a meal meant that, whatever the weather, we had to go outside to retrieve supplies for the next meal and to dispose of rubbish from the last.

Cynical though I was, thinking back to the good old days when we kept 20 man-days of food inside the tent, I had to admit that the constant availability of helicopters to go where we wanted represented progress of a welcome kind. It more than compensated for petty irritations in camp. Another bit of progress was

Darwin Glacier camp.

The next day, 14 November, was overcast with the sun faintly showing through, and it was still -10° F. I awoke after a snug sleep in a double sleeping bag. We had three helicopter visits in weather that, in my opinion, nobody should have flown in.

Now the survey started in earnest. While Radcliffe and Hughes flew out

The author surveying ice movement stakes on Byrd Glacier.

a solar (photo-voltaic) panel to power the radio. It seemed to work just as well in a whiteout as in sunshine. Mercifully, gone were the days when someone had to pedal furiously on a man-powered generator outside the tent. Gone were the days when the inside man had to tap away on a Morse key. Now we just spoke to the

over the glacier to position the fish floats, I followed them with some difficulty through the telescope of the theodolite. Other observers were doing the same thing from the upper and lower glacier camps. My position was on the summit of a 300-foot knoll; I had to stay there for three hours facing into a 20-knot wind.

Although wearing multi-layered clothing and two pairs of mittens, it was the coldest surveying I had ever done. Cheeks and fingers felt as if they were burning; later they lost all feeling. Bill King gallantly stood against me as a windbreak. During the day the glacier party put out six floats and then large black groundsheets to serve as targets visible in aerial photography. All returned safely, whereupon Terry and I entertained them with hot chocolate in our tent.

We needed to know whether the glacier itself was aground or afloat opposite our camp. Thirty miles to our right, the ice cascaded down from the polar plateau, so it was obvious that there it was aground. Fifteen miles to our left the tongue flowed into the Ross Ice Shelf, so there it had to be floating. There was no break in slope in the area between, nothing to suggest a sudden acceleration as the ice escaped from the drag of its rock bed. One way to find out would be to take successive vertical angles to the fish floats over a 12-hour period. If some rose and fell over the period, we would say that they were rising and falling with the tide.

A windy day followed, but we still had two visits from helicopters. By evening the wind was blowing 25 knots with drifting snow. Terry received some mail and zealously devoured it; I had to make do with reading a novel. We began the attempt to detect tides on 17 November, my 52nd birthday. It involved Terry and me alternately climbing the knoll to make hourly observations of the fish floats. Bill King and I took the first part of the night shift, Terry the second. It was miserably cold work, with our body temperatures reaching hypothermia level. Unfortunately, rapidly changing refraction made the results so imprecise that after 24 hours we were forced to declare the experiment a failure.

My diary says: "One helo visit today to deliver shopping." "Shopping" was a polite way of describing requests for food that we should have brought with us in the first place. In this case it was a packet of sugar. Terry liked brown sugar on his breakfast oatmeal. All we had was white sugar. So he asked for a two-pound bag of brown sugar to be delivered. To my astonishment, it was. John Doe — the famous taxpayer — remains blissfully unaware of how his money was spent.

The next day the temperature was so warm (0° F) that we were able to picnic in the open on top of the knoll. The glacier and the mountains provided the most spectacular back-drop of any picnic in my lifetime. Mark Hyland, who had joined us for the day, sat outside the tent reading a novel. At other times I have sunbathed naked in the Antarctic. We could have done it that day — though the slightest breath of wind sends one scurrying for shelter.

It was now time for the whole expedition to move upstream to new camps from which similar observations would be made. Terry and I went with the first helicopter and were supposedly put down at a site that Terry had selected earlier. Without looking around, we unloaded. Shortly afterwards, Terry realized that it was not the place he had chosen earlier. He called on the radio, but by that time the aircrew were eating lunch at the Darwin Glacier camp. We finally made contact with them by relaying a message through South Pole Station. Not unnaturally, the pilot, Lieutenant John Tennant, USN, was disgruntled at being asked for a second move.

The camp site we eventually reached was on a mountain summit 5,300 above sea level overlooking the glacier far below. It was an idyllic spot for helicopter-borne scientists but inaccessible, except to alpinists, by any other means. We managed to pinpoint all the survey markers by angles from at least two stations, and the distance between the camp sites was determined by tellurometer. More fish floats were set down and another attempt was made to detect tidal movement; but all to no avail.

We decided that the best solution would be to take simultaneous reciprocal vertical angles from camp and from the glacier. This should cancel out the effects of refraction. It meant camping on the glacier for three days — if we could find a safe spot. So, like a New York executive hailing a taxi, we summoned a helicopter; it came the following morning at 0940. Bill King and I loaded light-weight camp gear and dropped to the ice below. It took some searching before the pilots came upon a tiny bastion of bare ice surrounded by open crevasses. We landed in a hollow, the only place flat enough to pitch a tent. As the machine took off, the rotor blast was like a hurricane for the first few seconds. We lay on our rolled-up sleeping bags to stop them being blown into the nearest crevasse.

At synchronized hourly theodolite observations we saw the distant speck that represented one of our colleagues observing us from the top of the cliff. It was an eerie feeling to wave, and to see an answering wave, from a figure several miles away visible only in a 30-power telescope. Between observations we sat in the tent reading. Bill was immersed in Tolkien's *Silmarillion* and I in Solzhenitsin's *The First Circle*. It was Thanksgiving Day and our celebration consisted of dried beef stew, coffee, hot chocolate, and a slice of raisin bread.

We heard a helicopter on the third evening and guessed that it was looking for us — the proverbial needle in a haystack. They were searching in lazy circles several miles upstream because there was nothing to distinguish one bit of the glacier from another. Each time they faced our way I directed what I hoped would be a blinding flash at them with the signal mirror. It worked. My diary notes that the mirror "is certainly the most reliable homing device ever invented — if the sun is out." Since we were camped on the landing pad — the only level spot for miles — we rapidly packed up. The pilot, Lieutenant John Sullivan, USN, was impressed on finding us ready to go. "There are certain camps," he muttered darkly, "where people never seem to be ready."

We were returned to Darwin camp for "R&R" (rest and recreation). Finding out a bit more about the place, I learned that the power plant was kept running 24 hours a day. With its snow melter, the total fuel consumption was nearly 1,000 gallons daily, every drop of it airlifted here in LC-130 flights. A game of volleyball was in progress on the snow outside. The players were chased away from the helicopters after being reminded by the ground crew that each machine cost $1.5 million.

One day when fog rolled in, one of the

helicopters was caught away from camp. In visibility of less than 100 yards, the pilots felt their way home without radio aids, flying at a height of 20 feet. Pilots, I have noticed, have strong homing instincts when the alternative is a night in a survival tent.

The next trip was with an international aircrew. John Tennant had as his copilot Flying-Officer Denis Laird of the Royal New Zealand Air Force. This sort of collaboration has been common between the two countries; it did much to prevent misunderstandings which might otherwise have wrecked the good relations that prevailed between McMurdo Station and Scott Base, its neighbor over the hill. Back in our tent camp on the peak, the aircrew shut down their screaming engines and stayed for a cup of tea. Most of the aircrews refused to stop their engines in case there was a problem with re-starting. Terry Hughes was then airlifted down to the glacier to set out more markers. Supper was salmon steak, peas, and mashed potatoes, with hot chocolate to drink.

[1] Sydney Morning Herald, 18 September 1957. Quoted in: Elizabeth Chipman: *Women on the Ice. A History of Women in the Far South*. Melbourne University Press, 1986 (p. 86).

[2] Wally Herbert. *A World of Men*. London, Eyre and Spottiswoode, 1968.

[3] Jennie Darlington. *My Antarctic Honeymoon: A Year at the Bottom of the World*. Garden City, N. Y., Doubleday, 1956.

[4] Elizabeth Chipman. *Women on the Ice. A History of Women in the Far South*. Melbourne University Press, 1986.

[5] Paul Siple. *90° South. The Story of the American South Pole Conquest*. New York, G. P. Putnam's Sons, 1959 (p. 108).

[6] Sydney Telegraph, 5 March 1959. Quoted in: Elizabeth Chipman: *Women on the Ice. A History of Women in the Far South*. Melbourne University Press, 1986 (p. 87).

[7] Paul-Émile Victor. *Mes aventures polaires*. Paris, Éditions G. P., 1975. M. Victor died on 7 March 1995, aged 87.

[8] *Antarctic* (a news bulletin published quarterly by the New Zealand Antarctic Society), Vol. 4, No. 1, 1965, p. 14.

[9] Ibid., p. 16.

[10] Wally Herbert. *The Noose of Laurels*. London, Hodder & Stoughton, 1989 (p. 62).

[11] Walter Sullivan in *The New York Times*, 1 October 1969. Quoted in: Elizabeth Chipman: *Women on the Ice. A History of Women in the Far South*. Melbourne University Press, 1986 (p. 95).

Chapter Ten

The South Pole

A lot could be written on the delight of setting foot on rock.
It is like going ashore after a sea voyage.
R. F. Scott, 8 February 1912

On 28 November Tom Smith and I were flown to an unnamed and solitary peak at the head of the glacier. The temperature was +5° F and there was not a cloud in the sky. Landing on the summit was an adventure in itself. There was just room to touch the skids on the ground but not enough room for anyone to step out from either of the side doors. An incautious step would have had us tumbling over a cliff.

Come what may, alighting passengers are drilled to move forwards within the pilot's arc of vision. Any rearward exit could lead an absent-minded person to walk into the tail rotor. Faced with the choice of falling down a mountain or circumnavigating the tail rotor, I chose the latter. A frantic bellow from the crew chief confirmed that I had broken a cardinal rule. But it was my neck.

The peak overlooked a 50 x 50 mile area of cascading icefalls where tributary ice streams converge to drop into the Byrd Glacier canyon. It seemed like a canyon, though not many canyons on earth are like this one — twenty five miles wide. The rock was wind-sculptured sandstone intruded by diabase sills. We spent the next 10 hours taking rounds of angles and following the helicopter as it took Terry, Bill, and Peter out over the glacier to set up more markers. Henry Brecher was also following from another peak. We were all in radio contact. So distant was the helicopter that several times Henry lost it from view. The pilots were then asked to fly towards him until he latched on to them once again.

In these conditions I used the radio frequencies that work best. That day I was using 5,400 kilohertz, a Scott Base frequency, to keep in touch with Darwin camp. Nobody complained. Routine banter was taking place between various New Zealand field parties. At 1250 a new voice came in loud and clear, calling "Scott Base, New Zealand Nine Zero One." Scott Base replied, "Nine Zero One, I am unfamiliar with your call sign." This is a radio operator's polite way of saying "Who the hell are you?" With obvious relish, the stranger's voice continued "This is a DC-10 of Air New Zealand with 240 smiling faces and Sir Edmund Hillary on board." It was a sightseeing flight for tourists on a day trip from Auckland.

At the time, we were pleased to know that happy passengers were feasting on a bird's-eye view of our scenery. The significance of the little exchange over the radio only struck home 12 months later — exactly 12 months as it turned out. On 28 November 1979, at the same hour — 1250 — an Air New Zealand DC-10 with 257 people on board slammed into Mount Erebus in a whiteout. There were no survivors.[1] It was the fourth worst accident in aviation history. At an ensuing government enquiry it transpired that none of the three pilots on board had ever flown to Antarctica.[2] The airline had not bothered to transmit a flight plan to Scott Base, the only New Zealand outpost within 2,000 miles.[3] Another DC-10 pilot who had flown to Antarctica confessed: "I had thought that whiteout conditions were similar to fog."[4] Famous last words — as the saying goes.

We were day-trippers of a different sort. Left alone on a mountain top with the promise of being picked up at the end of the day, we had to be aware that things do not always go to plan. Our 10-page *Safety Guide* left nothing to chance:

> 1. Each USARP passenger is required to be dressed in the following cold weather apparel items before he will be permitted on any helo flight:
> Bunny boots or mukluks
> Long underwear bottoms
> Regular trousers (with wind pants preferred)
> USARP issue wool shirt
> Parka with hood (anorak not acceptable)
> Gloves or mittens with liners
> Balaclava (may be placed in the pocket during flight)
> The following additional items are strongly recommended to be worn or carried on person:
> Sunglasses
> Pocket knife
> P-38 can opener
> Spare pair of gloves or mittens
> Tops to long underwear
> Wind pants
> Ear brassiere
> Bear paws [heavy over-mittens]
> Two candy bars
> Small packet of nose tissue
> Approved ear plugs
> Signaling mirror
> Two clean handkerchiefs
> Four band aids
> Lunch
>
> 2. Each USARP, to be permitted aboard a helo flight, will also be required to have a "Lap Bag." This bag will be carried on one's lap while in flight or placed within the individual's personal reach without any assistance from another person while in flight. The Lap Bag will contain the following:
> Sleeping bag (single mummy)
> Biffy [bivouac] bag
> A pair of large sized wind pants
> Bear paws and spare pair of gloves
> Balaclava and scarf
> Can of survival rations
> Survival book
> 30 ft of parachute cord
> Matches in a sealed container
> Signaling mirror
> Pocket knife
>
> 3. A minimum of two passengers (no exceptions) may be placed in the field and the helo depart — providing these paxs have with them their Lap Bags plus a Survival

Bag. The contents of the "Basic Survival Bag Two-Man" are:
   Stove and fuel
   Survival food for two days
   Two-man tent
   Snow saw
   Folding shovel
   First aid kit
   Matches in a sealed container
   Signaling mirror
   Cook pot, cups, spoons
   50 ft of parachute cord
   Smoke bomb
   Four large candy bars

Nobody could fault the authors for lack of attention to detail. By my reckoning we now had three signaling mirrors. If that is a measure of their perceived importance in this sort of terrain, I will go along with it. On this day, the weather held, and we never looked inside the bags.

It was seven hours before they came to pick us up. We heard the deep-throated beat of the rotor blades half an hour before the helicopter came into view heading straight for us. That a machine 50 miles away could be clearly heard was an eloquent measure of the all-pervading silence of our surroundings.

Next day I went with Terry, Tad Pfeffer, and Peter Radcliffe to the left bank of the glacier to take ice samples for later physical analysis. John Sullivan was pilot, Denis Laird copilot. We put down between two parallel crevasses 20 feet apart and unloaded the ice drill. Believing that the helicopter was to wait on the ground while we drilled, Terry had not off-loaded his lap bag. Owing to some misunderstanding, the helo took off to go on another mission. We were left standing in the wilderness not knowing when or whether we would be picked up. The elaborate rules in the *Safety Guide* were only as good as the people applying them. Terry is a large man, so we had disagreeable visions of trying to share a sleeping bag with him. He too laughed at the thought.

In due course we were rescued and returned to Darwin camp. Lobster tails and filet steak for dinner. There followed more days of fish-float laying followed by survey from spectacular overlooks. In some parts of the glacier the surface was so broken up that we could not even touch down on one skid. The solution was to weight each fish float with a sack of rocks and hurl it from the door while hovering five feet off the ice. At windy survey stations we pitched the survival tent to brew up hot drinks. Mark Hyland and I were packing the survey gear one fine day when a helicopter came to pick us up. Not finding an area large enough to land on, they placed one skid on a rock while keeping the throttle at full blast. Mark fell on the rolled-up sleeping bags to stop them being blown over a cliff while I dived onto the flattened tent to prevent it following.

At some landing places the helo had to position itself quite precisely between large boulders. Some holes were found under the fuselage of one of the machines but none of the pilots would own up to having done the damage. Perhaps they were afraid of getting bad marks from the chief pilot, or they just had not felt the bump. In the circumstances, either was possible.

Thanks to Jim Fastook and his pocket calculator, the first results from our work began to appear on 2 December. One fish float was moving at the phenomenal rate of 2,822 feet per year. We had seen enormous rifts and confused crevasse patterns at the mouth of the glacier where it tore into the Ross Ice Shelf; now we understood why.

One of many visitors to our camp was Mark Leinmiller, a 19-year-old Boy Scout. When not scouting, Mark was a sophomore in mechanical engineering at Georgia Institute of Technology in Atlanta. Ever since Byrd's first expedition, American Antarctic expeditions have made a point of selecting a representative of the Boy Scouts of America to go south. The first, and perhaps best known, was Paul Siple, who wintered as a member of Byrd's expedition in 1929[5] and then made his career in polar research. Mark was this year's winner and was loving every minute of it.

There were days when we heard helicopters but never saw them. A Hercules came over at 26,000 feet to do the planned vertical photography, droning back and forth along parallel flight lines like a lawn mower. We spoke with them on the radio. Other Hercules daily left contrails high in the sky on their way to and from the South Pole. With all this activity we understood how it came about that USARP's budget that year was $131 million. Later we learned that Darwin camp and its associated activities had cost $13 million.

My time was up on 7 December. A Hercules landed at Darwin with a bunch of "DVs" (VIPs) on an inspection tour.

Terry Hughes deploying fish floats at the mouth of Byrd Glacier.

Among them was Jim Zumberge, my mentor from 1959 and now President of Southern Methodist University at Dallas, Texas. President Gerald Ford had appointed him to the National Science Board and it was in that capacity that he was appraising USARP's activities. The DVs came into one of the Jamesways to hear Terry and me, followed by the other "principal investigators" (as NSF grantees are called), expounding on the global significance of what we were doing. Some were impressed, others skeptical. It was their prerogative.

I was sorry to leave Terry and his ever-cheerful team of young scientists. Though most were half my age, I was never made to feel it. Taking off with the DVs, we flew north for an hour before parachuting a ton of supplies to a party led by Phillip R. Kyle, a geologist studying volcanic rocks. Back at McMurdo, I spent the evening imbibing with the DVs in their relatively luxurious quarters.

David Bresnahan, the NSF representative at McMurdo, took me on a 10-minute walk to Scott's *Discovery* expedition hut at the sensibly named Hut Point. The hut is normally locked to keep out souvenir-hunters but Bresnahan had the key. Successive groups of unpaid volunteers from New Zealand had been restoring the place but the interior was still a mess. A mess it should perhaps remain, because that is the way it was left by the last temporary occupants, the Ross Sea shore party of Shackleton's Imperial Trans-Antarctic Expedition in 1916. In the cold, dimly-lit interior, it was not hard to imagine the trials of life in the heroic age. The remains of dead seals lay on the floor where they had been butchered.

Shelves held canned food from the early years, labels stained by rust but many still legible: Huntley & Palmers "Digestive" and "Captain" biscuits, specially-made "Cabin" biscuits, Frank Cooper's "Oxford" marmalade, Fry's cocoa, Bird's baking powder, Coleman's mustard, Hunter's "Famed" oatmeal, and New Zealand corned mutton. There were the remains of clothing, boots, reindeer-skin sleeping bags, sledge parts, snow shovels, and a man-hauling harness. Shackleton's men had made an attempt with tarpaulins to partition off a space round the coal stove to trap a little of its heat inside.

In 1960 I had visited Shackleton's 1907–1909 expedition hut at Cape Royds, 20 miles north of McMurdo. It was smaller and much better preserved than the *Discovery* hut, presumably because it had served as a year-round home rather than temporary quarters. Scott's men had preferred to live in their ship moored 100 yards south of the *Discovery* hut.

I was invited to serve as a one-day field assistant to Tom and Davida Kellogg, a remarkable couple looking into the history of the pinnacled ice in McMurdo Sound. The Kelloggs, like Terry Hughes, hailed from the University of Maine at Orono. They already had one assistant, 26-year-old Carolyn LePage, a married M. Sc. student. Most of their work was done on day excursions by helicopter from McMurdo Station. The helo dropped us off on a frozen river between two 40-foot-high ice pinnacles. It was idyllic summer weather, sunny and warm with a temperature of +20° F. Left with survival bags only, we spent the day collecting shell fragments for radiocarbon dating, counting boulder sizes, and sieving soil in a search for microfossils. Having done this in one place, we marched for half an hour northwards to repeat the procedure. After four moves there was a pause for a sandwich lunch before continuing. When the helo returned late in the afternoon, the crew did not know where to search. Fortunately, they were flying straight out of the sun, so it was short work to attract their attention with the signal mirror. Having picked us up, the helo followed our footsteps — at 100 knots — back to the starting point to collect the survival bags.

The Kelloggs had been engaged in their painstaking work for years. Its outcome, after many vicissitudes, was a detailed history of glacier fluctuations in the McMurdo Sound area extending back thousands of years.[6] It turned out that the ice shelf itself was a comparatively young feature. Prior to 7,000 years ago, all of the floating ice shelf hereabouts was a grounded ice sheet resting on the seabed.

Life at McMurdo seemed more extravagant than ever, the diet more likely than ever to encourage obesity. Relief from the meal routine came on Sunday, when not only breakfast but also brunch was served. The menu for Sunday 10 December was:

Breakfast
Eggs and omelets to order
Bacon/ham slices
Pork sausage links
Creamed beef on hot biscuits
Home fried potatoes
Peach hotcakes with maple syrup

Brunch
Creole soup
Club sandwiches
Tuna salad sandwiches
Potato chips
Seasoned blackeye peas

Supper
French onion soup with Parmesan croutons
Steamship round
Baked perch fillets
Baked potatoes
Buttered succotash
Baked onions and tomatoes

In addition, there was a bewildering variety of drinks and an ice cream machine disgorging various flavors. If a post-prandial nap could be resisted, there were four movies to choose from, a lecture, and a closed-circuit TV film. They were shown in various clubs but USARPS were generally welcome at any. The menu for the same day, 10 December, was:

Officer's Club
    The Goodbye Girl — (PG) comedy
Chief Petty Officer's Club
    Flesh Gordon — (R) satire
Petty Officer's Club
    The Last Tycoon — (PG) drama
Enlisted Men's Club
    The Spiral Staircase — (PG) drama
Television
    Paco — (G) drama
Lecture
    "Richard Byrd in the exploration of Antarctica," by Lisle Rose, Polar Affairs Officer, State Department.

My remaining ambitions for the season were to visit an ice drilling camp on the Ross Ice Shelf, and to assess how the joint NSF/Scott Polar Research Institute radio-echo ice-sounding program had progressed in the 11 years since Gordon Robin and I had begun it back in 1967. In my experience, the NSF representatives at McMurdo were unfailingly helpful in trying to accommodate requests from scientists to make visits connected with their legitimate scientific interests, so I was able to achieve both objectives.

My route to the drilling camp 400 miles southeast of McMurdo was via the South Pole, 820 miles from McMurdo. It was aboard a Hercules carrying 10 ton of diesel fuel to power the station during the winter, and eight passengers. Another 17 ton of fuel was in the aircraft's tanks. The three-hour flight was in perfect weather, with an unfolding panorama of dazzling white mountains to our right. However, all the Hercules aircraft have pathetically small, round, and generally dirty windows, so photographic opportunities were

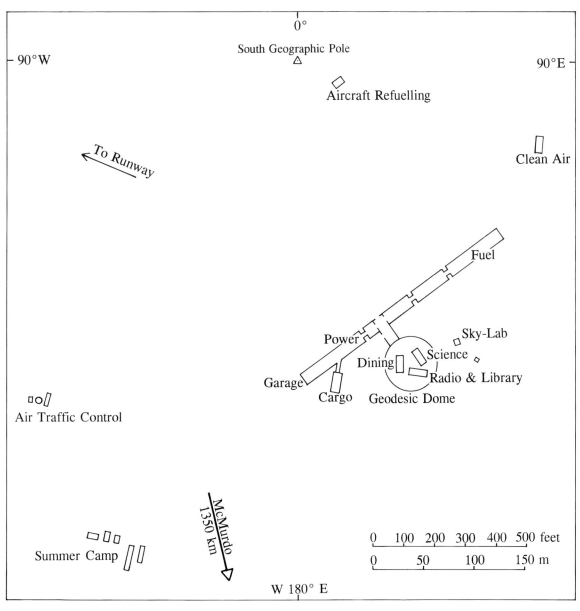

Amundsen-Scott South Pole Station.

limited.

Landing on a smooth skiway, our aircraft taxied in to a parking area festooned with pipes for carrying fuel to and from a vast undersnow fuel farm. Stepping off the plane, we were welcomed by Bill Benson, the resident NSF representative. A ramp led down to a snow-floored tunnel. Over the entrance hung a sign declaring

THE UNITED STATES WELCOMES YOU TO THE SOUTH POLE.

The tunnel led us into a giant aluminum geodesic dome, 164 feet in diameter and 52 feet high. Under the dome there are three two-story prefabricated huts. One houses laboratories and sleeping quarters, another contains the kitchen and dining room, and a third is for radio communications and a library.

The proper name of the place is Amundsen-Scott South Pole Station. It is the showpiece of the US Antarctic Program and I was delighted to discover that I could stay "overnight." In summer, this is the sunniest spot on Earth and often has 24-hour days of uninterrupted sunshine. Day and night are indistinguishable. The sun circles the sky at the same height above the horizon. Decisions about when to work and when to sleep are arbitrary. For convenience, the station keeps McMurdo time, which is also New Zealand time. Taken literally, a stay "overnight" could imply wintering here, because each night lasts six months — from 21 March to 21 September.

There were about 50 people in residence at this time of year, but only 18 for the coming winter. The youngest member of the wintering party was Andrew Cameron, whom I had known since his childhood. Andrew was the son of NSF's Program Manager for Glaciology, Dick Cameron, who had wintered at Wilkes Station in 1957. The sleeping quarters were cosy but also slightly claustrophobic, with windowless rooms holding two people on bunks, one above the other. There was a unisex bathroom to show the whole-heartedness of the recent switch from misogyny to equality of the sexes. Men and women were trusted not to peep behind the shower curtain to see who was inside.

The next day Bill Benson led a tour of the station. All but one of the wintering personnel were quartered under the dome. The exception was the doctor, Michele Raney, who lived in the "hospital" which was in a tunnel off the main entrance. She was to be the first woman to winter at the South Pole and the only woman in the wintering party. Michele had hardly recovered from the shock of being appointed to this most isolated post. On top of that, she had just received news that her mother had suffered a heart attack and was undergoing a high-risk operation for cancer. She seemed in two minds about resigning on the spot. However, having signed on to become the physician at the South Pole, she knew that resigning might be construed as letting the side down. Moreover, it would have been difficult to find another doctor at such short notice.

Michele showed me round her well-appointed hospital. Observing that she was under some stress, I casually asked whether it might not have been sensible to have at least two women on the wintering party. She rounded on me saying, "That would have been discrimination on the grounds of sex!" She continued, "I got my job here because I was the best applicant for it. My male colleagues got their jobs because they were the best applicants for their job." Suitably chastened, I quickly changed the subject.

Outside, the air temperature was -22° F, almost balmy by South Pole standards and evidently about average for the time of year. The annual mean temperature, I learned, was -56° F; the lowest ever recorded in winter was -117° F. Because I was well-dressed — in the Antarctic sense — the cold was no problem, but suddenly arriving at an altitude of 9,300 feet above sea level led to heavy breathing after exertion. However, I was able to walk the half mile to the original South Pole Station built in 1956.[7] Completely buried under snow, all that could be seen were a few radio masts and some rounded snowdrifts covering the taller buildings. By 1975, when the new station was built, access had become difficult and roofs had begun to cave in under the snow load.

The new station, with its geodesic dome, was carefully sited about 1,150 feet upstream from the geographic South Pole — towards which the ice (and with it the station) moves at a rate of about 33 feet per year. Once beyond the pole, the ice will be moving along the meridian of 40° west of Greenwich towards Filchner Ice Shelf and the Weddell Sea. Now, three years after the dome was built, it had begun to be buried under snowdrifts.

The "Clean Air" laboratory at the South Pole. The vertical pipes suck in air to measure the level of industrial pollutants carried by the wind from distant continents.

However, the rate of snow accumulation here is so low — about the same as rainfall in Death Valley — that the building could be usable for the next 20 years or so. Research programs under way included astronomy, aurora, radio-propagation, meteorology, geomagnetism, seismology, and biomedical studies. To my mind the most interesting work was in the "Clean Air" laboratory. This was a flat-roofed building with dimensions of 23 x 56 feet that was erected on stilts and kept 15 feet above the snow surface. Raised by periodic jacking, it was designed to stay clear of snowdrifts by allowing the wind to blow underneath.

A series of pipes projected 10 feet skywards from the roof. Fans pulled air from outside through instruments capable of measuring extremely low levels of impurities. Antarctica is known to be the cleanest place on Earth. Paradoxically, it also offers the best place in the world to study industrial pollution in the atmosphere, because whatever is found here represents global, rather than local, concentrations. Graphs on the walls of a laboratory within the building documented a steady rise in levels of carbon dioxide, methane, nitrous oxide, and halocarbon.

It was two days before the next plane arrived. It was a DV (VIP) flight carrying Graham Claytor, Secretary of the Navy; Norman Hackerman, Chairman of the National Science Board; and Richard Atkinson, Director of the National Science Foundation. They were led away by the staff for a grand inspection tour.

Evidently the Hercules crews are required to leave all four engines running, even if staying for an hour or more to have a meal. This eats into the fuel that otherwise could be given to the station. It is distressing to think of the cost. After lunching with the crew, I boarded the Hercules and took off at noon. The captain was the VXE-6 Operations Officer, Commander Pesce. Commander Morgan, the Commanding Officer of the squadron, was serving as navigator. I was pleased to note that although the aircraft was fitted with an Inertial Navigation System, the navigator did bubble-sextant observations of the sun just for practice. It has been a general rule that sophisticated navigation aids make life easier for pilots. However, they can go wrong, and when they do, only the old-fashioned methods can save the day.

We landed at the drilling camp in the middle of the Ross Ice Shelf; it bore the prosaic name "J-9." Two hundred miles to the southwest we could see peaks of the Transantarctic Mountains. The camp was a hive of activity. Only three hours earlier, three Russian glaciologists led by Igor Zotikov had finished taking a continuous ice core through the 1,380-foot thick ice, penetrating all the way to the sea beneath.[8] Understandably, they were elated.

John Clough, a glaciologist from the University of Wisconsin, led me to a freezer container and extracted the last section of the four-inch-diameter ice core. Twenty feet from the bottom, the ice that originally fell as snow on the surface gave way to frozen sea water, with its characteristic crystal structure.[9] The bottom face of the core consisted of the butt ends of large, vertically-oriented ice crystals, and there were what appeared to be algae inclusions. Glaciologists had wondered for years whether there was melting or freezing on the bottom of ice shelves. Here at last we had found evidence of freezing. There was no reason to believe, though, that it would apply everywhere.

The Russians had been brought in to obtain the ice core because of the failure of an American drill the season before. It was a wonderful example of scientific pragmatism. I admired the decision of NSF to swallow their pride and invite in the people most likely to succeed. Zotikov was a former Soviet exchange scientist, so had many friends in the west and was easy to work with. I had known him for years, so among other things, the day was a joyful reunion for both of us.

The American glaciologists had succeeded in a different way. Using a water-cooled flame-jet device, they had melted three large holes through to sea water. Two had been used to suspend instruments at various levels in the ice, while the third was for access to the sea beneath. It was 30 inches in diameter — quite big enough to fall into — and I approached the open top with caution because the snow surface was slippery. Directing a shaft of sunlight down the hole with a mirror, I could see the glint of sea water 200 feet below. So we were 200 feet above sea level. A lead-line sounding down the hole had shown that the sea floor was 770 feet beneath the bottom surface of the ice shelf. Sediment samples had already been recovered.

A small group of marine zoologists were having their own excitement. They had raised samples of sea water containing live, shrimp-like crustaceans. Elsewhere, these would be seen as whale-fodder; but no air-breathing mammal could survive this far under the ice. For the present, they were swimming in a glass jar in a domestic refrigerator in the Jamesway hut used as the camp mess hall. It must have been bewildering for them to be captured in the eternally dark, crushing pressures beneath the ice shelf, then to be held in a jar beside an electric light which came on whenever the door was opened. But they survived. Some were taken, still in good condition, back to the US.

J-9 was unique in another sense. It was the only Antarctic camp with an open-air swimming pool. Not that it was designed for recreation. The open-topped steel tank contained thousands of gallons of melted snow used to cool the flame-jet drill. The steaming surface was enticing to bathers. Nobody was swimming at the time but I was assured that it had been used. Just as I was contemplating a swim, the aircrew began rounding up the non-residents to be flown to McMurdo.

Or so we thought. On taking off, the Hercules headed due south. It turned out that two of the DVs had had enough of the Pole Station and wanted out. We were the best-placed aircraft to collect them. After refuelling at the pole, we flew the 820 miles to McMurdo in 2½ hours. Two helicopters were waiting to whisk the DVs to the base. Two helicopters to lift two people. Washington bureaucrats and Soviet apparatchiks have much in common. There was not even a snowmobile for ordinary mortals. We hitch-hiked on a passing truck.

The radio-echo Hercules was now ready for its first long flight of the season. The aircrew numbered seven. David Drewry, one of Gordon Robin's former pupils at the Scott Polar Research Institute, was supervising a six-strong scientific

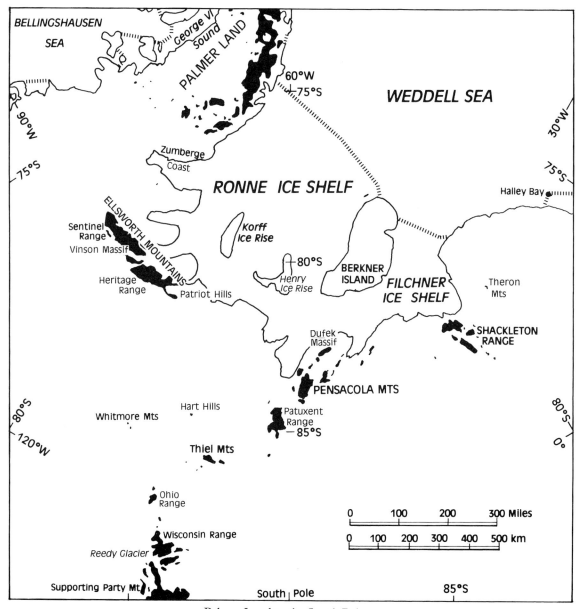

Palmer Land to the South Pole.

crew. It was the sixth season of NSF/SPRI collaboration. The scientific program had progressed a long way since our beginnings in 1967. Finn Søndergaard of the Technical University of Denmark had contributed an improved ice-sounding radar, and John Behrendt of the US Geological Survey had fitted a magnetometer to learn something of the rocks beneath the ice. The cabin was filled with a long row of instrument racks. The observers were arrayed before them, some peering into oscilloscopes, others tending chart recorders, and still others copying instrument readings into notebooks.

An extraordinary array of radar

antennas was mounted beneath the wings. The aerodynamic effects must have been horrendous but the major complaint of the pilots was that it slowed us down. We took off with 26 ton of fuel and cruised at 210 knots at an altitude of 22,000 feet. This was a bad height for ice-sounding but the best for fuel economy over the 1,200 miles between McMurdo and the target. The purpose of the flight was to study the subglacial extent of the Dufek intrusion, a bedrock formation dear to geologists. The only similar layered igneous intrusion of comparable extent is in South Africa, where it is known to contain valuable minerals. The Dufek Massif is a spectacular mountain range but it was believed to be only an outcrop of a much more extensive feature hidden beneath the ice. Behrendt's magnetometer could track its unusual magnetic "signature" as we flew along at 3,000 feet over the ice.

The aircraft ran into cloud over the most interesting area, but knowing the height of the terrain, and having radar scanning ahead and down, we could safely continue. The geophysicists became excited on discovering that the intrusion extends across Filchner Ice Shelf onto Berkner Island, much further than expected. We departed the area after a low level pass over the jagged peaks of Dufek Massif. Art Ford, a geologist who has made a special study of the area, was in camp on the west side of the range, and we saw his tents as the plane flashed by. By this time the Hercules needed refuelling. We landed at the South Pole after 9½ hours in the air. It was a glorious calm day, and felt warm at -22° F. After refuelling, it took another three hours to reach McMurdo.

The time had come to wend my way homeward. On 16 December, 57 returning scientists and support staff climbed into a Hercules for the flight to Christchurch. In spartan troop seats, like the rest of us, sat the Secretary of the Navy, the Chairman of the National Science Board, and the Director of the National Science Foundation. A guarded titter went round when a crewman revealed that the Secretary of the Navy usually had a personal Boeing 707 at his disposal.

Christchurch was 90 degrees warmer than the South Pole. Once we had dressed appropriately, it felt very little different. I shaved off my white beard and sunbathed to remove the "panda rings." All returning Antarcticans have the same problem: unsightly white patches where dark glasses have shielded the eyes from a suntan.

I was home by Christmas. Old timers of the Antarctic scene used to be scathing about those who came south for a five-week field season. We called them tourists. Now for the second time — 1967 was the first — I had joined them. Had my attitude changed? Well, yes. The difference nowadays is due to aircraft, to suffering almost no waiting time at McMurdo before being taken into the field, and to what is known as "remote sensing." The term includes radio-echo sounding, airborne magnetometry, and the use of satellite images to substitute for maps. We had been privileged to have access to all of these and to automatic instruments recording — in one hour — what earlier we would have been pleased to obtain in a month of ground travel.

I had not stayed to share in all that Terry Hughes and his team achieved by the end of the season, but I had helped them to start on the right track. The results appeared later. Back in 1960, my University of Michigan party had — with difficulty — measured the velocity at three points on Byrd Glacier. Now the survey team led by Henry Brecher had measured the exact position, elevation, and velocity at 600 points spread over the length and breadth of the glacier, an area of 1,500 square miles.[10] I breathed a sigh of relief on hearing that the new results were consistent with mine.

Reporting the figures was just the beginning. Terry and his colleagues next had to analyze them in terms of shear stress, strain rates, and the flow law of ice; to relate these to the history of the glacier; and to relate its history to the fluctuations and stability of the vast ice sheet that feeds the glacier. George Denton's team had established that about 18,000 years ago, the surface of Byrd Glacier had been up to 4,000 feet above its present level. This implied that, at the time, there could have been no Ross Ice Shelf but instead a much thicker ice sheet resting on rock.

Drewry's radio-echo sounding team had flown down the glacier with their remote sensing Hercules, so now we had an ice thickness profile over much of its length. In one area the ice was 9,000 feet thick. Cross sections we had made during the 1967 season showed that the ice is 3,000 feet thick opposite Horney Bluff. Now, 18 years after I began, we could report that Byrd Glacier discharges 4.3 cubic miles of ice into the Ross Ice Shelf every year. This staggering total is roughly

Pick-up at the end of a day's work on McMurdo Ice Shelf.

equal to the mass of snow which falls within its catchment area of about 400,000 square miles. It is also equal to the combined total discharge of all the other glaciers that feed the western margin of the Ross Ice Shelf.

A month after I left the Antarctic, Terry Hughes, Jim Fastook, and Tad Pfeffer were in a helicopter that crashed into a ridge during a whiteout. Tad suffered a fractured spine but the others on board were relatively unscathed. Terry later confessed that while the helicopter was rolling over and over and the cabin was filling with snow, he thought "I've made this trip somewhere before."

It was a flashback to his traumatic ride in a cement mixer at 15 years of age.

[1] P. T. Mahon. *Report of the Royal Commission to Inquire into the Crash on Mount Erebus, Antarctica of a DC-10 Aircraft Operated by Air New Zealand Limited.* Wellington, Government Printer, 1981.

[2] Air New Zealand McDonnell-Douglas DC10-30 ZK-NZP Ross Island, Antarctica, 28 November 1979. Aircraft Accident Report No. 79-139. Wellington, Ministry of Transport, 1981.

[3] Gordon Vette with John Macdonald. *Impact Erebus.* Auckland, Hodder & Stoughton, 1983.

[4] Stuart Macfarlane. *The Erebus Papers.* Auckland, Avon Press Ltd, 1991 (p. 202).

[5] Paul Siple. *A Boy Scout with Byrd.* New York, G. P. Putnam's Sons, 1931.

[6] Thomas B. Kellogg, Davida E. Kellogg, and Minze Stuiver. Late Quaternary history of the southwestern Ross Sea: Evidence from debris bands on the McMurdo Ice Shelf, Antarctica. American Geophysical Union, *Antarctic Research Series*, Vol. 50, 1990, pp. 25-56.

[7] Paul Siple. *90° South. The Story of the American South Pole Conquest.* New York, G. P. Putnam's Sons, 1959.

[8] I. A. Zotikov, V. S. Zagorodnov, and J. V. Raikovsky. Core drilling through Ross Ice Shelf. *Antarctic Journal of the United States,* Vol. 14, No. 5, 1979, pp. 63–64.

[9] I. A. Zotikov, V. S. Zagorodnov, and J. V. Raikovsky. Sea-ice on bottom of ice shelf. *Antarctic Journal of the United States,* Vol. 14, No. 5, 1979, pp. 65–66.

[10] H. H. Brecher. Surface velocity determination on large polar glaciers by aerial photogrammetry. *Annals of Glaciology,* Vol. 8, 1986, pp. 22-26.

Chapter Eleven

Runways on Ice

A man must make his opportunity, as oft as find it.
Francis Bacon (1561–1626)

A long and unexpected chain of events led me back to McMurdo 10 years after Terry Hughes' expedition. I had enjoyed five wonderful field seasons with USARP and never dreamed that there could be another. I had been involved with Antarctica for 39 years and was enjoying a quiet life at home.

It came about like this. Compulsory retirement age for employees of the British Antarctic Survey is 60, so in November 1986 I was cast to the winds — or a rocking chair — and given a pension. However, the rocking chair did not fit. While contemplating this problem, the telephone rang. It was Giles Kershaw, a professional pilot and former colleague in BAS. He and I had flown together in 1975 in the Antarctic Peninsula. That was Giles' first season, and from the start it became clear that he had a special talent for "bush" flying — if one can apply the term to operating in the Antarctic a thousand miles from the nearest bush. His phenomenal powers of concentration became evident when we flew a total of 214 hours together in six weeks in a DHC-6 Twin Otter, much of it at a height of only 30 feet over the snow in order to get the best results from radio-echo sounding. That record has not been matched since — at any altitude — by any pilot in Antarctica.

In the course of the next 15 years, Giles became the most accomplished pilot in Antarctic aviation history. Leaving BAS in 1979, he became free to fly non-government aircraft to a continent where only government aircraft were made welcome. The official line rested on the assumption that only governments had the means to operate safely. In the event of an accident, search and rescue operations could disrupt government activities. There were good reasons to be wary of accidents. Search and rescue activities following the DC-10 crash on Mount Erebus in 1979 severely disrupted the USARP program for that season.

In 1980, Giles was employed as a Twin Otter pilot in support of Sir Ranulph Fiennes' Trans-Globe expedition.[1] This was a private expedition that circumnavigated the globe via both poles. Although operating with only one aircraft, Giles arranged his own back-up options, and once turned the tables by saving the lives of three members of the South African government expedition who had strayed from their station without suitable equipment or experience.

When not flying in the Antarctic, Giles flew in Greenland and made many landings on pack ice in the Arctic Ocean using the Tri-Turbo, a DC-3 fitted with three engines instead of two. This aircraft was unique in the true sense — only one was ever built. It was modified from a standard DC-3 by replacing its relatively unreliable piston engines with modern turbine power plants of approximately the same horsepower. So impressed was Giles with the performance of this hybrid machine that he chartered it from the owners, POLAIR of California, and flew it to the Antarctic in 1984 carrying a group of climbers to Vinson Massif in the Ellsworth Mountains, at 16,080 feet the highest mountain in Antarctica.

When the Tri-Turbo was no longer

Midnight in Palmer Land.

available, he continued to take climbers into the Ellsworth Mountains with Twin Otters under the banner of Adventure Network International (ANI), a Canadian company set up for the purpose. The flying arm of ANI became the world's first commercial airline in the Antarctic.[2]

But, to return to the telephone call in November 1986, Giles and I had known since 1975 about bare ice, or "blue ice," areas in the Ellsworth Mountains. Although irrelevant to our activities at the time, it occurred to us that the ice might be smooth enough to take conventional wheeled aircraft, avoiding the need for skis. Now he reminded me of what we had seen and asked my opinion. I said it could only be judged by careful examination on the ground (in this case, the ice). He thereupon invited me to come and survey the place.

Although more than 99 percent of the surface of the Antarctic ice sheet consists of snow, the existence of "blue icefields" was well known.[3] They occur in places where wind and sublimation (or evaporation) strip off the snow which covered the ice before it flowed into a localized area of mountain turbulence. However, the process can be very slow indeed. Some aluminum survey markers that I myself drilled into the surface of a blue icefield in 1951 were found still standing 34 years later. Changes in the exposed length had averaged less than one inch per year.[4]

On 4 December 1986 Giles and I flew an ANI Twin Otter from a base camp at Vinson Massif to reconnoiter sites in the Heritage Range that we had glimpsed at a distance 12 years before. After finding an extensive blue icefield at Patriot Hills (80° 19' S, 81° 20' W), we made low passes to inspect the surface. Then came the acid test: I asked Giles to land. It took some minutes for the hydraulic system to raise the skis so that the wheels could project beneath. We were apprehensive on the final approach. On touchdown I felt a high-frequency vibration as the wheels ran over a wind-scalloped surface that we had seen from the air. The bumps were only about two inches high, thus easily absorbed by the low-pressure tires.

I spent five days camped on the site with Canadian geophysicist Mike Maxwell. We determined its surface slope and roughness by using an ordinary engineer's level and staff. There was a potential 10,000-foot runway which appeared adequate for transport aircraft. The big problem was turbulence caused by migrating vortices coming over a ridge on the upwind side. Almost all the snow falling on the ice had been swept away. Later we discovered where it had gone. It had been dumped a mile or two downwind — in areas where the turbulence had moderated.

I became convinced that, in the long run, bare ice landing strips would benefit government operators as much as any other. But it was clear that no government would commit transport aircraft to landing on icefields until the concept was proven. We were reminded that throughout Antarctic history, most logistic innovations have been the work of private expeditions. Governments prefer to follow well trodden paths.

On the strength of my report, ANI set out to prove the concept by using a Douglas DC-4, a four-engined airliner without skis, the following season. The US Government tried hard to stop the operation on the grounds that it was too hazardous and would probably involve the US in search and rescue. The US ambassador in Santiago, Chile, asked the head of the Chilean Air Force (Fuerza Aérea de Chile, or FACH for short), General Fernando Matthei, to refuse permission for the aircraft to proceed to the Antarctic. Under Chilean law the general had power to do this.

Fortunately I had been introduced to General Matthei on the way home the previous season. He was a member of the

Chilean junta under General Pinochet. Pinochet was getting a bad press outside Chile — notably in the USA — but I found Matthei to be a gentle, charming, and highly intelligent man. He spoke fluent English and was interested in the possibility of taking his own wheeled aircraft to our icefield. It was obviously in his interest that we — and not FACH — should bear the risk of a proving flight. I never found out what words he used to give the US ambassador the brush-off.

The American dislike of mountaineering expeditions seemed to be focused on foreign expeditions. In 1966–1967 there had been an American Antarctic Mountaineering Expedition that was not only allowed to climb in the Ellsworth Mountains but also was flown there in an LC-130 at taxpayers' expense. The 10 members were selected by the American Alpine Club to ensure that the first ascent of the continent's highest peaks could be credited to Americans. Members of the expedition attained the summits of five peaks including Vinson Massif, the highest point on the continent.[5]

Although people think of the Antarctic as a hazardous place, we who work there know that some major hazards have to be overcome before anyone sets foot on the ice. Having dealt with the political obstacles, the inaugural flight of our DC-4 from Punta Arenas to Patriot Hills on 17 November 1987 was relatively uneventful, though it took 11 hours 43 minutes to cover the 2,000 miles. The captain of the aircraft was Jim Smith, and as we approached for landing, he summoned me to sit in the jump seat between him and his copilot. The implication was clear. If the landing ended in disaster, he wanted me to share the blame.

The icefield was not an empty spot in the wilderness. The crews of two ANI Twin Otters and two FACH Twin Otters were lined up to witness the landing. The Chilean Twin Otters had come all the way from Marsh, a Chilean base 1,500 miles away, presumably at the behest of General Matthei. We landed straight towards the ridge, an approach that is normally against any pilot's instinct. But the wind was blowing over the ridge at 35 knots, so Jim knew that he could stop in time. Touchdown was as gentle as it had been the year before. It was my 61st birthday and it felt good to be back on the ice, despite an air temperature of -13° F.

In the years since, there have been well over a hundred non-stop flights to the icefield. The distances are such that each take-off of the DC-4 from Punta Arenas had to carry not only a useful load but also return fuel to allow for 24 hours in the air. Although flying in the Antarctic is unlikely ever to match the safety record of the world's major airlines, ANI can match the record of government operators.

Returning home with the second DC-4 flight in 1987, I wondered whether my friends in Washington would now regard me as another *bête noire* of Antarctic aviation. Giles had defied the US Antarctic establishment on a number of occasions, so he already enjoyed that distinction. I need not have worried. The ice runway concept now proven — ANI having taken the risks — the logistics staff at the National Science Foundation rapidly came to espouse the idea of using icefields for their own purposes. I knew that they had funded — a decade earlier — a study of some icefields in the

Pensacola Mountains with a view to using them for wheeled aircraft,[6] but then never followed up the idea. Although snow-free bare ice covers only one percent of the area of the Antarctic ice sheet, the total area amounts to some 40,000 square miles. I had little doubt that other icefields would in due course be developed to handle inter-continental traffic.

(CRREL) in Hanover, New Hampshire. Malcolm had been involved with Antarctic glaciology for more than thirty years, was a pilot himself, and had become leader of CRREL's Antarctic engineering projects.

The immediate reason for NSF's interest was their proposed rebuilding of the South Pole Station sometime before

First DC-4 landing at Patriot Hills blue ice runway, 22 November 1987.

Having opposed ANI's activities, the US Antarctic Program now needed someone to pursue the idea on their behalf. In a typically American act of magnanimity, they came to me. Well, I like to think it was they. The actual invitation came from Malcolm Mellor, an English *emigré* employed by the US Army Corps of Engineers at their Cold Regions Research and Engineering Laboratory

the year 2000. The present station had been occupied since 1975 and — like all buildings on the ice sheet — was progressively becoming buried under snow. Supplying the station entirely by air — using LC-130 aircraft on skis — was becoming expensive. These ageing aircraft were limited to payloads that, in any commercial airline, would be considered uneconomic. Using the same method to bring in possibly 4,000 ton of cargo and fuel for a new station would be so costly that they would face a stark choice — either give up the idea, or sacrifice the scientific program for some years to devote all resources to the rebuild.

The alternative possibility — if an ice runway could be found within a few hundred miles — would be to fly in all bulk cargo with large wheeled aircraft and then carry it to the South Pole by tractor train. Some modern jet transports could lift in one flight all that an LC-130 could carry in 20 flights. The question, therefore, was "Are there any icefields within a few hundred miles of the South Pole that could offer a smooth enough and long enough airstrip for transport aircraft?"

As usual, success would depend on meticulous and unhurried preparation. The horrendous cost of flying in the Antarctic could only be justified if I knew exactly where to search. So I went to the US Geological Survey (USGS) in Reston, Virginia. As the principal mapping agency of the US Government, USGS had accumulated an archive of more than 300,000 aerial photographs of Antarctica obtained, over a period of 40 years, for mapping purposes. About 7,000 of these covered the area of interest for the project. Thumbing through them, I identified 37 sites worth reconnoitering from low-flying aircraft. However, the photographs were taken from a height of 20,000 feet and it was impossible to estimate surface slope or to see small crevasses. It was clear from the start that most sites might prove unusable on grounds of slope, grade change, length, crevasses, or obstructed approaches. We could only find the good ones by looking at all of them.

The author.

My life has brought a bountiful supply of interesting questions that can only be answered through field work. Here was yet another. When I first went south in 1949 at the age of 22, I would have said that Antarctic exploration was a game for young men. It would be crazy to allow anyone over 60 — "with one foot in the grave" — to get near the place.

But times have changed and so have I. So, it seems, have the people who decide these things. Now it was 1988, and at age 62, I had no doubt that I was up to the task. I was asked what was needed to do the job. For the sixth time in my 30 years' association with USARP, I had the privilege of answering that question, knowing that everything possible would be done to provide the means. I submitted this list:

> One Twin Otter aircraft dedicated to the task for a month.
> 9,000 gallons of aircraft fuel at the South Pole.
> Prints of 70 of the aerial mapping photographs.
> Camping gear for a party of four.
> Trail food for 150 man-days.
> One small snowmobile.
> Surveying equipment.
> One small sledge.

That was easy. Then came the impedimenta of bureaucracy. CRREL instructed me to fill in a host of forms. They were to be sent to ITT Antarctic Services, Inc., a contractor to NSF responsible for handling the formalities. A letter dated 26 October 1988 acknowledged my papers in these terms:

> *Dear Mr Swithinbank:*
> *Thank you for your recent inquiry concerning employment with ITT Antarctic Services, Inc.*
> *We have carefully reviewed your qualifications and although they are excellent, we find that we currently do not have a suitable position available to discuss with you . . .*
> *Very truly yours,*
> *ITT Antarctic Services, Inc.*
> [signed] *Mr. Joseph A. Greco*

It was clearly a standard letter, so the qualifications of every rejected candidate were considered "excellent." At least I could not take offence. It was evident that the left hand did not know what the right hand was doing. It was the last I heard from ITT. Malcolm Mellor sent me an airline ticket to New Zealand.

Reaching Christchurch on 30 November, I learned that USARP was now called USAP (US Antarctic Program), though some old-timers still addressed the scientists as USARPS. The deletion of "Research" from the name of the enterprise was said to be to appease the civilian contractor's staff, whose connections with the research end of the operation were tenuous. In the view of most scientists this was a mistake that might give the wrong signal to foreign governments about the legitimacy of US operations in Antarctica. If it was not dedicated to research, what was it? But the New Zealanders followed suit: NZARP became NZAP.

The following day brought the familiar routine of trying on polar clothing. Much has changed over the years because, for most Antarcticans, outdoor activities now involve little exertion or exposure to high winds. The result is that really windproof outer clothing has been replaced by a heavily insulated parka that is not particularly wind-proof. Since my work has always been in the open, miles from any shelter, I bring my own wind-proof outer garments. Parkas are for sedentary workers or for standing around in.

At 0700 next day I was again at the clothing store on the way to check-in for the flight to McMurdo. Others were converging on the place, some young, some old, some bored, some apprehensive. The only one I recognized was Harry Keys, a New Zealander studying the distribution of icebergs in the Ross Sea. USAPS are required to don their polar clothing, or at least some of it, in the 70° F heat of the Christchurch summer. "Civilized" clothing is then left in the clothing store to await our return. Some people leave too much in the store, and rue the day on finding that some buildings at McMurdo are kept at the temperature of a Christchurch summer.

On checking-in at the joint USAP/NZAP passenger terminal, a sniffer dog sniffed at each piece of baggage in turn. It was not some New Zealand authority looking for a bomb but instead the US Navy on the lookout for drugs. I felt very sad that our pristine wilderness could ever be sullied by people needing delusion to escape from it. Why couldn't they have stayed home? But times have changed and so have the people who come south. I was with scientists as dedicated as any in the past, but also perhaps some escapers from debt or from women. Or women escaping from men. I did not dwell on it.

After a long wait, we were herded into a Royal New Zealand Air Force Hercules. RNZAF and NSF do a lot of sharing of logistics in the interests of economy. Occasionally it has led to bickering about who gains and who loses, but common sense generally prevails. It will be a sad day for Antarcticans when accountants impose their own tunnel vision on the give-and-take decades that my generation has enjoyed. We were eight USAPS and eight NZAPS; today, at least, the accountants would be happy. The aircraft was identical with the NSF kind except that it had no skis. The US Government has never allowed the Lockheed Corporation to sell the LC-130 wheel/ski version to anyone outside the US. Someone in the Pentagon must have a reason — or a whim.

After 7½ hours we touched down on the sea-ice runway. The sun was shining and the vista was as grand as ever. To the north, the volcano Mount Erebus with its plume of steam. To the south, White Island (which is white), Black Island (which is black), and Brown Peninsula (which is brown). Beyond was Minna Bluff. Who was Minna, I pondered, and whom did she bluff?[7] In the southwest there was Mount Dromedary, Cape Chocolate, Beacon Heights, and Mount Morning. Beyond them, the serried peaks of the Transantarctic Mountains, some reaching to 13,000 feet.

A bus carried us to the station, bumping across the tide-crack by Scott's *Discovery* expedition hut, now in quite good condition. The "Kiwis," as the New Zealanders are known here, were taken off to Scott Base in their own minibus. Although some clean four-story accommodation blocks had replaced the dingy slums I lived in 29 years ago, McMurdo is still a frontier settlement and unattractive. We were driven to the NSF Chalet, sat down and addressed by Ron La Count, Senior US representative in Antarctica, and David Bresnahan, NSF representative. They read something akin to the Riot Act — rules of behavior never to be broken. I had heard these 10 years before, but now there was a dire and formal warning against helping "private expeditions." At first I thought it was an allusion to my own infamy of last year in helping ANI to fly into Patriot Hills. But no! It was Greenpeace.

Greenpeace had established a wintering station beside Scott's hut at Cape Evans, and kept observers in tents near Scott Base to observe everything that was going on, there and at McMurdo. They had been needling the American station for some years, mostly about waste management. La Count made an inexplicit reference to "events earlier this season." I learned later that graffiti had appeared one night on some fuel tanks. The wording was such that Greenpeace became the prime suspect. But there could be no proof. It might have been a USAP.

Following the rather negative tone of the "welcome briefing" we were handed room keys for the Mammoth Mountain Inn. The next morning I was summoned to a meeting in the Chalet, to be confronted by 10 officials with Ron La Count in the Chair. To them, the agenda was unwritten yet implicit, but it took me some minutes to understand this. Their serious mien made me feel that I was in the dock.

I had the wrong end of the stick. They had been summoned to find out what I needed for the project. I was dumbfounded. In my earlier incarnations at McMurdo it had been a case of finding out where to go and who to ask, then endlessly chivying until either you got what was needed or realized that you must

McMurdo in 1991. Two ships are in Winter Quarters Bay. Five Navy and two Coast Guard helicopters await their next assignment.

do without. Now someone must have seen the light — or employed a management consultant — or been threatened with the sack. However it came about, the organization had reached a heavenly pinnacle of maturity with scientists near the summit.

La Count asked "When could you be ready?" I replied "This afternoon." My claim was grossly unfair and unrealistic; it produced a stunned silence. It was Saturday, I was told, and nothing gets done at weekends. Keeping a five-day week, I realized, could have reduced my life's work in the Antarctic by almost one third. McMurdo was evolving from a frontier outpost to a permanent settlement and I had to expect the same working hours as back home.

I was glad to discover that not everyone kept the weekend sacrosanct. Walking to the Berg Field Center, I found the Manager, Ann Peoples, hard at work. She was extremely helpful, providing a new engineer's level to determine the slope of our icefields, and a vast list of camping equipment to select from. Then on to Deb Baldwin, Manager of the USAP food store, a wisp of a girl who was a top-class white-water canoeist when not in the realms of ice. She offered a list of foodstuffs that could grace Aladdin's cave. Finally to the USAP garage, where Bruce Winter checked out a skidoo, complete with spare parts and instruction manual.

At this stage I was CRREL's sole representative on the ice runway project. It did not take long before people were doing everything possible to ingratiate themselves in the hope of being asked to serve as my field assistants. Several who had come to the Antarctic on the strength of their wilderness experience had found themselves confined within the largest settlement in Antarctica. The prospect of camping high in the Transantarctic Mountains made me an object of envy, but I was not ready to sign on a crew before finding an icefield worth surveying.

Sunday was quieter than Saturday. I climbed Observation Hill, a 750-foot pyramid on the edge of town, which gives a fine view in all directions. On the summit there is a cross erected by the survivors of Scott's last expedition as a memorial to the five members of the polar party who perished on their return from the South Pole in March 1912. It is nine feet tall and made of Australian eucalyptus. On it are carved the words:

IN
MEMORIAM
Capt. R. F. Scott, R. N.
Dr. E. A. Wilson, Capt. L. E. G. Oates, Ins. Drgs., Lt. H. R. Bowers, R. I. M.
Petty Officer E. Evans, R. N.
WHO DIED ON THEIR
RETURN FROM THE
POLE. MARCH
1912
TO STRIVE, TO SEEK,
TO FIND,
AND NOT TO
YIELD

The lines were chosen by Apsley Cherry-Garrard from Tennyson's *Ulysses*. Some years ago the Antarctic Treaty designated the site a "Historic Monument" and erected signs in English, French, Spanish, and Russian to record the fact. The English language sign had been stolen, and there were cigarette ends and discarded film wrappers lying about. It was a measure of the number of visitors to the site and the lack of historical perspective on the part of a few.

Professor William (Bill) Cassidy of the University of Pittsburgh gave a lecture on meteorites in the evening. Almost all the scientific staff attended, also about 100 of the support staff — an impressive turn-out. It was partly that Bill Cassidy was congenial and known to many of the locals because he came every summer, but also that people were excited at the thought of fragments from space or other planets raining down on ours. More meteorites — or meteorite fragments — have been found in Antarctica than on any other continent. A few are believed to have come from the Moon,[8] some perhaps from lava flows on Mars.[9] Most were discovered on blue icefields such as those I had come to study. I agreed to keep a look-out.

McMurdo sprang to life on Monday morning. There was a meeting with Captain John Smith, Commanding Officer of VXE-6 squadron, and A. J. Brown, the ITT/Antarctic Services Manager. Chief Petty Officer Pankey issued me with a radio and showed how to use it. As I had decided to base the flying on the Amundsen-Scott South Pole Station, I arranged with Rik Campbell and Steve Marvin to fly everything there.

There were two Twin Otters supporting USAP operations. They were chartered from Kenn Borek Air, of Calgary, Alberta, and flown by Canadian aircrews. Operating independently from VXE-6, I sensed some resentment among the Navy aviators at the Canadians' freedom. One of them described Twin Otter pilots as "cowboys." Having flown the season before with the same company, I knew that the verdict was unjust. In fact most of the Canadians had more polar flying experience than the Navy pilots. I was delighted to learn that one of the pilots of my aircraft was to be Brydon Knibbs, copilot on the DC-4 flight from Punta Arenas to Patriot Hills twelve months before. I spoke with him on the radio and

In 1960, we were housed in canvas-covered Jamesway huts. Thirty five years later, most people live in two- or three-story accommodation blocks.

we agreed to rendezvous at the South Pole three days later.

Having assembled everything necessary, it all had to be packed according to very strict rules. Hazardous cargo, such as white gasoline for the camp cooker, had to go on an LC-130 flight with no passengers. Robin Kinnaird and his colleagues in the USAP cargo warehouse turned to with vigor and everything was made easy. I

RNR, who in 1902 commanded the relief ship *Morning* on Scott's *Discovery* expedition.

All now being ready, I flew on 8 December to the South Pole as sole passenger on an LC-130. Met by the Station Manager, William (Bill) Coughran, I was given a berth in a double room next door to where I had stayed 10 years before. On the surface, little had changed,

Twin Otter enroute to the South Pole. The pilot has an astro-compass mounted on the window-sill to check his heading.

thought back to the way things were done in 1960 and welcomed the change. At dinner that evening I met Jack Renirie, NSF Press Officer, whom I knew from the 1960s; and Julian Ridley, a young field assistant who confided that he was a grandson of Lieutenant William Colbeck,

and it was easy to find my way around. Now there were 86 people in a station built for a summer population of 40. The temperature outside was -27° F. In the galley it was +70° F.

The Twin Otter arrived the next day from Siple Coast. In commuter service at home, the same type carries up to 20 passengers, but not here. The cabin was encumbered with two cylindrical fuel tanks, each holding 250 gallons. These were made necessary by the enormous distances involved in ferrying the aircraft from its home in Calgary. Even within Antarctica, it had to operate over distances for which it was not designed. In addition, a substantial weight of survival equipment had to be on board, and it was augmented in proportion to the number of passengers.

The aircrew consisted of Brydon Knibbs, my friend from the DC-4 last season, and Brian McKinley — both of them qualified captains. The engineer was Tony Frigon. All three were Canadian with thousands of hours of polar flying under their belts — I was in good hands. They treated their expensive machine with the respect it deserved. Although the engines, even at the South Pole, would start from cold just with the on-board battery, their lifetime between overhauls could be greatly extended by being pampered. Accordingly, the crew hitched a cable from the main diesel power station and drew a constant six kilowatts to keep both the engines and instruments warm. Another reason was to keep the gyro-compass in their inertial navigation system (INS) running day and night. If inadvertently stopped at the South Pole, an inertial system will not re-align with the Earth's axis until it is taken a considerable distance northwards.

Bill Coughran convened an afternoon meeting with the pilots and Bob Hurtig, "facilities engineer." I explained how I proposed to operate, and what help was needed from the station. They promptly agreed to everything without argument, and the meeting was over in half an hour. I could not help comparing this with the series of committee meetings that the British would have required before reaching any decision. Throughout my stay at the South Pole, people were unfailingly helpful. Although never discussed, it was evident that very careful personnel selection was the order of the day, more so than at McMurdo.

The altitude of the South Pole presents no problem for healthy people. Aircraft, however, require a much longer take-off run than at sea level. A snow runway just over 2½ miles in length is kept "groomed" with a grader.

That same afternoon, an LC-130 arrived from McMurdo. Shortly before it landed, there was an announcement over the PA system: "X-ray-delta-zero-two due in at 1630 with 38 unwanted visitors." This neatly summed up the attitude of most isolated stations to "intrusions" from outside. At the same time, everyone knew that the station was totally dependent on supplies brought in from outside by the "intruders."

Being the showpiece of the Antarctic program, however, South Pole does suffer a surfeit of visitors: Congressmen, government officials, and journalists. Today the journalists were being chaperoned by Jack Renirie. Necessary though such visits may be, they do disrupt the scientific activities for some hours. Senior scientists are required to show off their work and to justify it in relation to the enormous costs of keeping the station running.

Boarding his aircraft for the return flight to McMurdo, one portly gentleman caught sight of our Twin Otter and exclaimed, "What's that cute little plane doing here?" Cute or not, our machine did look very small beside the Hercules.

After supper in the smallish dining room, I invited Brydon and Brian to share half a bottle of Bailey's liqueur while we yarned about old times — in this case, last year. Liquor is freely available for purchase from the station shop at duty-free prices, and considerable quantities are consumed. We were warned, however, that in the rarefied air of altitude it takes less alcohol to unsteady the balance.

The next morning, 10 December, I was raring to go. The weather was sunny, clear, and calm, with a temperature of -26° F. After checking weather-satellite images of the Transantarctic Mountains and loading fuel, survival gear, maps, sandwiches and vacuum flasks of hot chocolate, we were airborne at 0955. The flight plan was to go straight to the nearest blue icefield that I had identified from the aerial mapping photographs, look it over, and then continue — in the most economical order — to others.

It would be flippant to report that we headed north — all directions from the South Pole are north. Directions have to specify a meridian of longitude. We headed north along 150° W and soon saw mountains ahead. On a clear day, mountains can be seen from the pole by climbing to just a few thousand feet. It took only 70 minutes at the Twin Otter's sedate speed of 140 knots to reach Mount Howe, the world's southernmost nunatak.

Mount Howe is at the head of Scott Glacier, on which Tom Taylor, Art Rundle and I had camped 27 years before. It consists of a long nunatak with — on its lee side — four square miles of moraine in ridges aligned with the direction of ice flow. Beyond it, there is a vast area of blue ice with dimensions of about one by five miles. This is what we had come to see.

We let down to a height of 30 feet and flew from one end of the icefield to the other, timing the run with a stop-watch and recording the change in altimeter readings from one end to the other. From this I was able to calculate that the average slope of the ice was acceptable for a potential runway; it was also remarkably smooth. We were heartened because this icefield, being close to the South Pole, was the most favored.

But it might not be the best. So we flew on to three more sites, all of which were found unsatisfactory for one reason or another. At this point bad weather loomed ahead, so we returned to Mount Howe, raised the skis above the wheels, and landed on the bare ice without difficulty. Whereas we could have landed with the skis down, the ice would have scratched their friction-reducing plastic coating.

We stopped the engines and set out to inspect the place. Reaching the moraine after walking half a mile, we found that it contained everything from giant boulders to fine sand. The smoothness of the ice was encouraging. We must come back here for a more thorough investigation.

Returning to the South Pole, I felt vindicated by our early success. However, while the dimensions of the ice were adequate and the slope was within published runway criteria, we knew that to most pilots, the very idea of operating wheeled aircraft in the interior of Antarctica was still seen as heresy.

[1] Ranulph Fiennes. *To the Ends of the Earth.* London, Hodder & Stoughton, 1983.

[2] Charles Swithinbank. Antarctic Airways: Antarctica's first commercial airline. *Polar Record,* Vol. 24, No. 151, 1988, pp. 313-316.

[3] Valter Schytt. Blue ice-fields, moraine features and glacier fluctuations. *Norwegian-British-Swedish Antarctic Expedition 1949–52, Scientific Results,* Vol. 4E. Oslo, Norsk Polarinstitutt, 1960.

[4] Karsten Brunk and Rudolf Staiger. Nachmessungen an Pegeln auf einem Blaueisfeld im Borgmassiv, Neuschwabenland, Antarktis. *Polarforschung,* Vol. 56, No. 1/2, 1986, pp. 23–32.

[5] Samuel C. Silverstein. The American Antarctic Mountaineering Expedition. *Antarctic Journal of the United States,* Vol. 2, No. 2, 1967, pp. 48–50.

[6] A. Kovacs and G. Abele. *Runway site survey, Pensacola Mountains, Antarctica.* Hanover, New Hampshire: US Army Cold Regions Research and Engineering Laboratory (CRREL Special Report 77-14), 1977.

[7] Minna Bluff was named for Minna, the wife of Sir Clements Markham, Captain Scott's mentor in planning the British *Discovery* expedition of 1901–04. A bluff is a headland with a broad precipitous face.

[8] O. Eugster. History of meteorites from the Moon collected in Antarctica. *Science,* Vol. 245, No. 4923, 1989, pp. 1197–1202.

[9] D. D. Bogard, L. E. Nyquist, and P. Johnson. Noble gas contents of shergottites and implications for the Martian origin of SCN meteorites. *Geochim. Cosmochim. Acta,* Vol. 48, 1984, pp. 1723–1739.

Chapter Twelve

Envoys from Outer Space

> *Living mainly on horse meat now, and on the march, to cool our throats when pulling in the hot sun, we chew some raw frozen meat.*
>
> E. H. Shackleton, 1 December 1908

Next day it was Brian's turn to occupy the left-hand seat. We headed for Reedy Glacier, named after Rear Admiral James Reedy, USN, who took over command of the Naval Support Force when Tyree retired in 1962. Two hours later we met dense cloud. Climbing to 14,000 feet in an attempt to cross the Wisconsin Range, we could see that the way ahead was barred. So Brian set course once more for the pole.

Having time on our hands in the afternoon, I walked half a mile to an outlying station established to house three separate groups of cosmologists. On the way I was overtaken by several cross-country skiers, rapidly skating their way along the track just to keep fit. Ah, youth! For the first time, it made me feel old.

The scientists were researching into cosmic background radiation, which is believed to be the remnant of the primeval fireball from which the present-day universe evolved. So attenuated is this by the time it reaches our planet that the slightest radio or electrical interference compromises the signal — hence the need to get away from the main station. The three groups were collaborating by studying different aspects of their science, but at the same time — it seemed to me — each competing for the big breakthrough that might redound to their credit. Special large antennas faced skywards. It was a big enterprise that must have cost millions of dollars to establish, not to mention at least 100 hours of Hercules flying time. Mine was not the most expensive project that season.

My diary for the day ends unhappily: "Got message that our aircraft is needed at McMurdo — curses on them!" So much for an aircraft dedicated to the ice runway project. However, flexibility being the key to mental equilibrium in our business, my words never got beyond the diary. Ron La Count had asked to use our aircraft for a week — and he was the boss. So we fuelled the wing tanks, the main tanks, and an extra 350 gallons in the cabin. I could have stayed behind at the pole but chose to go with the aircraft to keep pressure on for its return. We took a mile to get airborne, leaving 1½ miles of runway unused — a comforting safety margin.

We diverted from a straight path to McMurdo to take a look at possible ice runway sites near the head of Beardmore Glacier. My flight log records one of them

Unloading at midnight.

as "Too soft and not long enough," another as "Stepped profile but probably sufficient length on a step." Climbing to 16,000 feet to clear mountains in our path, Brydon glanced back at me and asked over the intercom, "You OK, Charles?" It was a fair question. We both knew that supplemental oxygen is recommended at that altitude — particularly for old men. But I felt fine.

It took 6½ hours to reach McMurdo.

At dinner I met Richard Otto and Simon Stephenson, both of whom used to work in my department at BAS. They belonged to a NASA team studying the Siple Coast — with its great ice streams that I had seen in 1961 from the nose cone of the ill-fated P2V. It was good to know that ever since we began, there had been continuity and a logical progression in studies of the glaciers that feed the Ross Ice Shelf.

The next morning I went to see Ron La Count, obliquely hinting that I would be around until the Twin Otter was returned. I think he was expecting an angry reaction to his hijacking. However, I knew there was nothing to be gained by impetuosity — he was used to dealing with it. Relieved, he reached into his desk drawer and withdrew a bottle of wine. We understood each other.

I met more old hands. Ed Zeller and Gisela Dreschhoff who, like me, seem to have been around since before the dawn of time; and Jerry Kooyman, a seal physiologist who was at McMurdo in the sixties. Then Tommy Foster, the present Navy liaison officer at NSF. Another physiologist, Bob Elsner from Alaska, is married to one of my wife's college friends. Bob knew Ove Wilson, the Swedish doctor on my first Antarctic expedition (1949–1952). It was a far cry from the early days, when most polar explorers had one opportunity in a lifetime to go south. They lived on nostalgia for the rest of their days.

I have done some low flying in my time, but the next "flight" was the lowest ever. There was a hovercraft on trial, and it had been used with some success over level surfaces. It was a Hake "Husky" G1500TD designed to carry 10 passengers or 2,000 pounds of cargo.

I had been asked to drill some shallow holes on the *Pegasus* crash site seven miles southwest of McMurdo, to see whether the snow was saturated with melt-water. *Pegasus* was a Super Constellation that crashed many years ago; the remains were where they had come to rest. Malcolm Mellor had proposed to develop the site for year-round operation by wheeled aircraft. In preparation for the flight, I was handed a pamphlet *Hovercrafts and You*. The pilots were Louise Czarniecki and Sarah Jones. Brydon Knibbs and Tommy Foster came along to help. Though I had once met the inventor of hovercraft, Sir Christopher Cockerell, and ridden across the English Channel on the world's largest, this was a novel experience.

We flew a foot off the ground at a speed of 18 knots. Directional control was comparable with that of an aircraft without a rudder, and the only way to stop was by doing what Brydon would call an intentional ground-loop, skidding through 180 degrees so that the thrust direction was reversed. Neither of these characteristics prevented the craft from being a useful form of transport, faster and more comfortable than normal oversnow vehicles. We stopped in several places for drilling to a depth of three feet, finding no water in the holes. Malcolm Mellor had hoped for dry snow, and it was.

Since there was no other task for the machine that day, we headed for the edge of the fast ice — where an unbroken sheet of sea-ice meets the moving pack ice. Without a hovercraft, this would be a dangerous place to be. Several lives have been lost when fast ice broke up and carried people off to sea. Now we stopped the engine and had a picnic. A procession of Adélie penguins strutted by to see what all the noise had been about.

It was 19 December before I got the Twin Otter back. Heading south, we again sought to economize on flying time by looking at blue ice sites on the way. One of them — Lewis Cliff Ice Tongue — was also the site of a big camp of meteorite hunters led by Bill Cassidy,[1] the man who had lectured to us at McMurdo two weeks before. The camp held eight people in four tents, neatly pitched in a row. We "buzzed" them at low level, they waved, and we went our way. We were at the South Pole 6¼ hours after departing McMurdo.

By now we had examined 14 out of the 27 possible ice runway sites that I had selected from aerial photographs. The best so far was on Mill Glacier, a tributary of the Beardmore. Mount Howe was the second best.

The next day we flew to inspect blue icefields at Mount Pool (crevassed), Colorado Glacier (sloping and bumpy), Reedy Glacier (ideal, but too far from South Pole), Berry Peaks (undulating), Mount Nichols (melt holes), Harold Byrd Mountains (soft and cracked), Scott Glacier (too much turbulence for a close inspection), Mount Salisbury (moraine on surface), and Koerwitz Glacier (too much slope). None of these obstacles had shown up in the aerial photographs. It was a day to remember — weaving our way among immense glaciers and skirting mountain ramparts of surpassing grandeur. I found I could project my mind into the utter silence of the landscape below, becoming oblivious to the roar of the engines.

It was becoming clear that we had found no place as good as Mount Howe within reasonable striking distance of the South Pole. It was time to have a closer look at it, camping there for some days. We had the camping equipment but nobody had agreed to accompany me. Discussing this with the aircrew on the way home, they surprised me by offering to serve as the camp crew.

This had the immense advantage that air transport would be on hand. I had already decided that whoever came camping, it would not be fair to ask them to forego Christmas celebrations at the South Pole. Now we could return to the pole whenever the survey was completed.

Next morning we loaded the camp gear and took off. Engineer Tony Frigon, who normally stayed at the pole station keeping an ear to the radio, came with us in case there was any problem with the aircraft. We had been given two pyramid tents, each designed for three people. While it was nice to have the space, they presented a greater surface area exposed to the wind, which was potentially dangerous in the turbulent air that we expected. Moreover, they had been badly rigged by new people at McMurdo who evidently had never pitched a pyramid tent in a blizzard.

This illustrates one of the few drawbacks of the modern method of putting people into the field. In earlier times we had to rig and test our own tents — there was nobody to do it for us. Like a parachutist who packs his own canopy, we could blame only ourselves if something went wrong. I remembered my cold first night with Terry Hughes on Byrd Glacier, having relied on others to select the sleeping bags.

My companions had carefully planned how to secure the aircraft. We knew that Twin Otters, even with their wings, nose, and tail firmly tied to the ground, had been flipped on their backs and destroyed in turbulent winds. That was because the direction from which the winds attacked them was unexpected. Here we had to plan for extreme winds from any direction.

Oil tankers loading at sea had solved this problem, using what is known as a single-point mooring. The tanker is secured to a buoy by a single bow-line and oil can be loaded over the bow. The ship rides to the wind or waves, from whatever direction they come. We used the same method. Drilling a hole in the ice with a hand drill, a steel stake was frozen in place with ice-water. Tony led heavy chains from the stake to strong points on the nose of the aircraft.

It is terrifying to be inside any aircraft when a high wind changes direction. I knew a man who did it. Once in a lifetime, he told me, was quite enough. The structure creaks, groans, and rocks from side to side as the wind tries to lift a wing. Finally, with an agonizing lurch, the machine skids round its anchor and realigns with the wind. But it survives.

Brydon Knibbs, Twin Otter pilot.

I shared a tent with Brydon; Brian and Tony shared the other. By the time we had secured everything and made a meal, it was time for bed. Next morning the wind was about 15 knots with gusts to 20 knots. The temperature was 0° F. The plan decided on was to mark out what seemed to be the optimum long runway with a line of bamboo poles drilled one or two feet into the ice. Then we would work along it with the engineer's level, making offsets to determine the lateral gradient.

We had brought the skidoo and sledge, on which Tony and Brydon now loaded the ice drill and a large bundle of stakes. They set off on a compass bearing of 090°, planting a stake every few hundred yards. After lunch they went in the opposite direction — 270° — while I re-rigged our tent to make it safer. The camp radio failed us, so we checked in with Pole station using the aircraft's transmitter.

Two more shopping days to Christmas. But although it was clear and sunny, 23 December brought a stinging 20- to 30-knot wind. Work outside was impossible. We read books, yarned, and slept.

The wind dropped at 0500 on Christmas Eve. I bellowed reveille to the other tent but there was not even a grunt in reply. Brydon and I were out by 0515 but had to wait half an hour for the others. Now we could start the survey. We went to the north end of the "runway" and began "leveling" southwards. Brian went ahead with the 13-foot measuring staff, I sighted on it through the telescope of the level, and Brydon "booked" — writing the numbers as I read them off the staff. We "leapfrogged" past Brian and sighted backwards before signalling him to leapfrog past us. This is the conventional way to compensate for refraction, which would otherwise distort the profile.

For each move, I lay on the sled, known as a "banana" sled because of its rounded form, while Brydon towed it behind the skidoo. The object was not to save me from walking but to speed the operation. I held the delicate instrument on my stomach, which served as a shock absorber. Without the cushioning effect of pneumatic tires — as with the aircraft — driving over the corrugated surface was a bone-shaking experience for both man and machine.

Surveying was cold work, but we made good progress, and by the end of a long morning's work had completed more than four miles of leveling. This length is more than adequate for the largest aircraft in the world. The best part of it was a 10,000 foot stretch of ice with a mean gradient of only 0.3 percent. There were some small bumps — not more than a foot high — which would need planing off with a grader.

We had skipped breakfast, so now stopped for lunch. Afterwards, we surveyed a line at right angles to the long line. It was level and 6,460 feet in length, but the shorter length would — in service as a runway — be offset by its alignment into the prevailing wind. Finally, walking across to the moraine, we filled a bag with rocks to take to the South Pole. People who spend a whole year out of sight of land — in the ordinary sense — relish a tangible reminder that Antarctica is made of rock, even though it may be a mile or two beneath you or a few hundred miles over the horizon.

We also filled a bag with ice chipped from the surface with an ice axe, knowing that drinks laced with the real thing taste better. Meanwhile Tony had been preparing the aircraft for departure. We took off at 1820 and landed at the pole in time to join a pizza party. Later, I interrupted a Christmas Eve celebration in the galley to present everyone with a hunk of rock and a chip of ice.

Christmas dinner was the result of fine collaboration between professionals and amateurs. Steve Midlam and Alan Bronston led the galley crew. They were assisted by Elsie Gowdy, an ever-smiling grandmother who loved working at the pole. We "subsisted" on:

> Whole roasted fresh New Zealand pig
> Roasted hindquarters of venison with poivrade sauce
> Rice pilaf with sautéed onions
> Baked apples a la Raoul
> Fresh sautéed vegetables
> Herb Rolls
> Pumpkin pie
> Pecan pie
> Stollen-Christmas loaf
> and an unlimited supply of wine and fresh fruit.

Amundsen must have turned in his grave — this was the South Pole! We continued drinking and yarning late into the night.

The next day was R&R. I decided that we needed more surveying at Mount Howe, and that it was unfair to tie up the aircrew where they could not fly, in case a need arose elsewhere. Except for Stephen (Denny) DenHartog, a CRREL glaciologist who was on his way south, I had declined — on grounds of economy — Malcolm's offer of assistants. I had a hunch that there might be another way. Bill Coughran suggested putting a sign-up notice in the galley to see if there might be volunteers. When I looked at the board some hours later, there were 25 signatures on the paper. I needed three. Anxious to avoid any suggestion of bias, Bill picked three names out of a hat.

One was Douglas Chichester, a geophysicist from the Geological Survey who was also the station scientific leader. Then Kay Tate, who seemed to do almost anything at the station, and John Gushwa, a "materialsperson" in the quaint language of the contractors. All of them were keen to go on any terms. But first they had to arrange for others to cover their work — a condition that Bill had made part of the deal. I never found out what inducement they offered, but pinch-hitters were recruited in very short order.

I sought one luxury — an outhouse. Answering nature's call under the open sky — or in blizzards — has never been a favorite part of the Antarctic experience. With the persistent winds that we encountered on the icefield, it became trying. So I sought the advice of Ray Brudie, the station carpenter. He took it as a challenge — to design, find materials for, and build the outhouse in less than a day. The result — by Antarctic standards — was one of consummate luxury — a prefabricated building designed for easy on-site erection by unskilled workers, complete with seat, sloping roof, and lockable door. It could be disassembled and transported to any new site.

The rest of the party set about gathering other equipment that we needed. The excitement of going camping was infectious, and at one point I counted eight people helping us towards an early start. They were not disappointed. Outhouse and all, we were airborne at 1143 on 27 December.

The weather was sunny and clear, and the newcomers were struck by the beauty

Kay Tate reading while ice chips are melting on the stove.

of the Mount Howe scenery. Coming from the featureless south polar plateau that was their home made the experience even more of a delight than it was for me. I tented with Kay, and Doug was with John. After a quick lunch, Doug set off to number the runway markers and measure their height above the ice surface. In this way, a second measurement a year later would show the

next job was to make close-spaced leveling profiles to test whether any short-wave roughness of the ice surface was within published limits for heavy wheeled aircraft. For the most part it was, so we were encouraged. I also measured the approach and climb-out angles that are of critical interest to pilots.

Walking across to the moraine, Doug

Discovery of the world's southernmost meteorite (foreground) at Mount Howe.
From left: John Gushwa, Doug Chichester, Kay Tate, and the author.

rate at which the surface was being eroded. It was too windy for survey work, so the rest of us organized everything for an early start the next day.

After chatting on the radio with the South Pole over breakfast, we set off to survey lines perpendicular to the long runway so that aircraft could land even when the wind was blowing across it. The

and John returned with a heavy, rounded piece of rock that they guessed was possibly a meteorite. The ice near the moraine was strewn with wind-blown chips of rock, but this one had caught John's eye as being different. It was oval in shape, with no sharp corners, and had a rough surface. Its weight and size and slightly-reddish dark-brown color suggested a content of iron. The surface texture resembled densely spaced pin holes. Realizing that this could be the southernmost meteorite ever found, and a whole one at that (most are fragments), we were electrified to hold in our hands this messenger from outer space. None of us had ever come close to a meteorite before. Here was something from a million miles — or light-years — beyond our ken.

Holding it in our hands, I remembered, was the last thing we should be doing. The chemistry of meteorites can hold clues to their origin — to the composition of the parent body. Experts analyse them to search for amino acids, possible traces of life on other planets. Handling had ensured that at least the outside of this one was covered in amino acids. But there was no restraining our excitement as we discussed what to do with this all-time souvenir of our camp at the bottom of the world. We wanted to keep it, but at the same time we guessed that it was a treasure that should be examined by experts. Resolving to offer it to Bill Cassidy, who was searching for meteorites 300 miles away, we thought of a compromise. We would give it to him and at the same time ask that when the meteorite was finally cut up for analysis in some ultra-clean laboratory back home, four tiny thin slices should be cut so that each of us could cherish the memory of its discovery. But if a name had to be associated with it, that name should be John Gushwa — the finder.

The next morning we spoke with Pole Station to ask for a pick-up when convenient. Doug and John went off to measure the remaining stakes while Kay and I strode off towards the moraine, each of us secretly hoping to collect another meteorite. Though we must have walked past several, the moraine consisted of rocks of all sizes, making it impossible — without the eyes of an expert — to identify them. There were boulders the size of a cottage, boulders the size of a car, boulders the size for a rock-garden, heavy cobbles, and sand — in places almost quicksand because the sun had melted some of the ice. It was strenuous walking but full of interest.

We sat down on a rock for a snack. Kay remarked on the exhilaration that came from hiking where no person had ever hiked before. She had two grown children back in Montana, but was revelling in our shared experience in this distant place. We were far from home and our homes were far from each other, but we had in common a love of wilderness places, the wilder the better.

Our daydreams were shattered by the sound of an aircraft. It must be ours. We had become so absorbed by the magic of our surroundings that neither of us had looked at a watch. The plane buzzed our camp and we guessed their surprise at seeing nobody emerge from a tent. Evidently our colleagues were still out measuring. The plane landed but it was an hour before they noticed — with some relief — the four of us converging on the camp from different directions. Brydon was well on the way to folding the tents, so we were soon off the ground and heading for the pole.

Not soon enough, it transpired. Since the Otter had been on the ground longer than expected, and her radio was silent, McMurdo had diverted a Hercules to check on us. In our view, the reaction was hasty, but it was a measure of the change

The author at Amundsen-Scott South Pole Station with Kay Tate and Elsie Gowdy.

from the 1960s, when McMurdo didn't begin to worry unless our radio was silent for a week.

At the pole, I carried our treasure — the meteorite — to the scales in the galley. Elsie Gowdy announced that it weighed 5½ pounds. We had showers and read letters from home. At dinner that night, I passed the meteorite round the table. We did not have the courage to deprive them of the obvious pleasure. Now 80 pairs of hands had contaminated it. Should we admit this to Bill Cassidy? Or should we do as the professionals do — seal it in a sterile plastic bag and pretend that nobody had touched it?

Despicably, I chose deceit, consigned the thing to a clean plastic bag, and tied off the opening. How would they ever know?

Bad weather kept us on the ground the next day, so I rested, checked the camp gear, and sought volunteers for the next camp. My last camp crew had worked like Trojans and cheerfully done everything asked of them, were unfazed by cutting winds and clearly had loved the experience. All of them volunteered to come on the next camping trip but I wanted to be even-handed. I was being besieged by other volunteers.

The last day of 1988 saw the arrival of two LC-130s. In one was Peter Wilkniss, head of the Division of Polar Programs at NSF; Jan Faiks, a state senator from Alaska; and Denny Hartog who had come to take part in the search for ice runways. Wilkniss' Hercules sat on the ground with all four engines running for two hours while he enjoyed a leasurely lunch and a tour of the station. At lunch I sat next to him and was pleased to note that he knew who I was and what we were doing. It turned out that it was on his initiative that the ice runway project had been approved.

He made clear that the pole Station was here to stay, but it would need replacing with a new structure before the end of the century.

I can find no adjective to describe New Year's Eve dinner.

Amundsen-Scott South Pole Station
New Year's Eve 1988
Dinner Menu

Appetizer

Seafood Sausage
A homemade seafood sausage containing gulf shrimp, rock lobster tails, sea scallops and filet of sole freshly poached in court bouillon served with a light tomato cream sauce.

Spanekopeta
A creamy combination of spinach and three cheeses wrapped in filo dough puffed to a golden brown in the oven and served with a tomato herb sauce.

Soup

Vichyssoise
A French favorite — creamy smooth potato onion soup served chilled with a dash of chopped chives.

Gazpacho
A Spanish classic — zesty tomato and cucumber soup flavored with a selection of herbs served chilled with croutons and a touch of sour cream.

Salad

A marinated assortment of hearts of palm, sliced green peppers, artichoke hearts, avocado slices and capers, decoratively arranged dressed with a tarragon vinaigrette.

Intermezzo

Citrus Sorbet
A tart refreshing lemon-lime ice churn — frozen for smoothness — the perfect way to lighten the palate for the entree.

Entree

Filet Mignon Steffon
A freshly cut mignon char-broiled to perfection stuffed with a shiitake mushroom duxelle, wrapped with bacon, topped with a touch of pesto served with a Madeira demi glace.

Chicken Cordon Bleu
A chicken breast filet is rolled around a slice of smoked ham and provolone cheese, lightly breaded and pan fried to a golden brown served with a creamy veloute sauce.

Le Veghead Platter
A decorative dish containing a tangy cheese fondue en croute surrounded by sliced eggplant, toasted almond pasta, broccoli timbale, glazed carrots, stuffed tomato provencale, pommes au gratin and sliced beets.

Vegetables

Pommes Champignoise
Fresh small potatoes cut into mushroom shapes sauteed in garlic butter and parsley.

Pommes au Gratin
Thinly sliced potatoes baked in a creamy cheddar sauce topped with grated parmesan.

Broccoli Timbale
Steamed broccoli baked in a savory custard served in a muffin shape garnished with strips of fresh carrot pasta.

Glazed Carrots
Fresh carrots are tourneed and steamed, then sauteed with brown sugar and butter.

Bread

A freshly baked combination cloverleaf roll consisting of wheat, rye and white breads topped with kosher salt and served with butter curls.

Dessert

South Pole Bombe
An ice cream cake in the shape of a dome consisting of a vanilla sponge cake — chocolate butter cream roll layered with homemade Kahlua coffee ice cream topped with pralines and white chocolate-covered espresso beans, then surrounded by a Baileys hard sauce.

We were served by volunteer waiters and waitresses dressed for the occasion. Red wine, white wine, or Champagne went with every course, and for those who had any space left, there was coffee, liqueurs, and sweetmeats. My admiration for the dedication, professional skill, and sheer hard work of the galley staff knew no bounds.

Now Scott too must have turned in his grave — this was the *South Pole!*

---

[1] William A. Cassidy. Meteorite search at Lewis Cliff ice tongue: Systematic recovery program completed. *Antarctic Journal of the United States,* Vol. 24, No. 5, 1989, p. 44.

Chapter Thirteen

The Acid Test

It's not getting to the pole that counts. It's what you learn of scientific value on the way. Plus the fact that you get there and back without being killed.
                    Richard E. Byrd (1888–1957)

Denny Hartog roused the aircrew on New Year's Day 1989 and we headed for Beardmore Glacier to reconnoiter more icefields and also to drop into Bill Cassidy's place with the meteorite. It took 2½ hours to reach the meteorite hunters' camp. They were not expecting us, but swarmed out of their tents when they heard the Otter. Taxying up to the group, we found a welcoming-committee of eight beaming faces. Cassidy stepped into the Otter to get out of the biting wind, and we excitedly thrust the package into his hand. Holding it, and before even looking at it, he declared it to be an iron meteorite. No other meteorite of that size would weigh as much.

We were wondering how careful he would be — whether he would even open our far-from-sterile bag. It was with immense relief that we watched him put his bare hand into the bag and extract the treasure. We enquired about precautions against contamination. He explained that iron meteorites are most unlikely to contain amino acids and need not be kept in sterile conditions. Absolved of any temptation to deceive, I breathed a sigh of relief and confessed all.

Seeing Cassidy's interest, I offered to carry two of his party to Mount Howe to search for more. I had no authority to divert our aircraft from the icefield search but decided against asking. For years I had been shocked by the tunnel-vision of scientists who refused to lift a finger on behalf of any branch of science other than their own. Believing that their work was more important that anyone else's, they passed up opportunities that — with minimal sacrifice — might vastly increase the productivity of the research enterprise taken as a whole. Here was an opportunity to practice what I preached.

Cassidy could not release anyone at once, so we flew off to Snakeskin Glacier to search for a fuel cache that had been set out for us by an LC-130 on skis. We had earlier selected a smooth, level, and crevasse-free site, and notified VXE-6 of the map coordinates. We found the cache some distance away on a three percent slope in an area of large sastrugi. We could only conclude that because our site had

First landing on Mill Glacier, 2 January 1989. The medial moraine in the distance separates Mill Glacier from Beardmore Glacier beyond.

been selected by Twin Otter pilots, the advice had been ignored. There were two 500-gallon rubber "sealdrums" on 9- x 7-foot aluminum pallets. The LC-130 crew could have saved $10,000 by simply rolling the sealdrums onto the snow and recovering the pallets for reuse. This would have taken only a couple of minutes, but evidently that was considered too much. We decided that after using the fuel, we would at least rescue the sealdrums. The pallets were too wide to fit in the Twin Otter.

Snakeskin Glacier is 330 miles from the South Pole and we needed the fuel to get home. The Twin Otter is a priceless asset in terms of being able to land almost anywhere but we were stretching its range.

The cache was needed for safely operating in this, the most distant area of interest in the search for ice runways.

Using the Snakeskin cache, we spent some hours investigating icefields from the Shackleton Glacier in the east to Beardmore in the west. The best site of all was on Mill Glacier off Plunket Point. The point was named by Shackleton in 1908 to honor the Governor of New Zealand. Shackleton mapped it during his exhausting climb up the Beardmore to reach the polar plateau. Here on Mill Glacier was the very same patch of blue ice that I had seen from a single-engined Otter 27 years before. We now landed on it — wheels down — and were at once impressed by its smoothness and vast extent. After walking around for a time, we concluded that this had to be the next camp site.

Since Cassidy's people were now ready, we dropped in to pick up two of his team and their snowmobile. John Schutt and Ralph Harvey had been chosen to make a thorough meteorite search of the Mount Howe icefield. Two hours later, on our flight home, we landed them at a food cache that had been left at our old camp site. By the time we ourselves reached the pole, the populace were fast asleep. We had spent nine hours in the air since taking off 11 hours before. Insomnia was never a problem.

Bad weather the following day gave time to prepare for camping. Denny Hartog, for one, needed no guidance. A glaciologist in his mid-fifties, he had wintered at Little America V 30 years ago. I was offered as field assistants Valerie Sloan (materialsperson) and Mark Parent (meteorologist); both were in their twenties. Anxious to avoid any hint of a generation gap, I chose to tent with Valerie. Denny and Mark would share the second tent.

At 0920 on 4 January we took off for Mill Glacier. Mill is the biggest tributary of the Beardmore and offers a virtually crevasse-free route to the polar plateau. Had Shackleton known about it, he could have shaved miles off the route that he did

The confluence of Beardmore Glacier (left) and Mill Glacier (right) as seen from Plunket Point.
The round-topped mountain is The Cloudmaker, 56 miles away. Ninety miles from us,
the peak of Mount Kyffin near the mouth of the glacier just shows above the horizon.
The landing threshold of our ice runway is behind Denny Hartog's head.

take. Landing where we had landed before, we pitched camp, planning to spend a week on a thorough study of the area. Instead of leaving the aircrew idle for a week, I felt we should offer the Twin Otter to a glaciological party working 400 miles away on the opposite (east) side of the Ross Ice Shelf. They had one aircraft already but they could use a second. Their party included my former colleagues from Cambridge who had been at McMurdo a couple of weeks before. Nationality may inhibit collaboration in some sciences — but not here.

Setting out survey lines only needed two of us, so Mark and Valerie walked over to explore the left bank of the glacier less than a mile away. Here was a vast area of boulder-covered ice-free land known as the Meyer Desert, named after George Meyer, a microbiologist who worked in the vicinity in 1961 and who, like me, had once been an exchange scientist with the Soviet Antarctic Expedition.

With crevasses to the east of us and land to the west, it was evident that our putative runway had to be directed up-glacier. This had the advantage that it would be facing into down-glacier winds, probably the most common wind direction. A mile down-glacier there was a high medial moraine marking the junction between Mill and Beardmore glaciers.

We had brought with us the tiny snowmobile that we had used at Mount Howe. Denny and I set off upstream with a small ice-drill and a bundle of bamboo poles to mark a proposed survey line. One of us drove while the other had a bone-shaking ride on the small sledge towed behind. By the time we were stopped by crevasses, we had traversed four miles.

Adding that distance to the stretch below our camp, we had found a runway long enough to land a space shuttle. Spacecraft were not, however, in our plans, so all we had to do was find the best one or two mile stretch for conventional aircraft. On the way back to camp we drilled the stakes in a dead straight line to guide the survey.

Chili for dinner with wine that Mark had hidden inside his sleeping bag. Valerie and I decided that we would play cook on alternate days. Since the cook had to select ingredients from the odd assortment of foods that now weighed down the tent flap, it meant that we could enjoy our favorite food on alternate days. This worked well. My brief diary entry ends:

Wind a biting 15 knots all day but temperature +10° F, balmy. I used a face mask driving into wind. Magnificent view down Beardmore as far as Cloudmaker. We are a happy bunch.

Peering out of the tent at 0700 the next morning, I found the wind blowing at 25 knots, so retreated inside. We called South Pole and passed them our weather observations. There was a slight improvement at noon, so we dressed up and emerged to face the elements. Waiting for really good weather could take a long time. Lenticular clouds over the mountains indicated high winds aloft but we could work if we kept moving.

Valerie and Mark were invigorated by their walk yesterday, so they now set off in a different direction while Denny and I worked down-glacier to finish setting up the stake line.

Standing on the massive medial moraine to look out over the Beardmore, we spied four dark-colored objects on the ice some distance away. Working our way across, we found four empty jet-fuel drums dated 1960. Evidently these had been left by a helicopter involved with "Topo-South," a major topographic survey operation that took place here in 1961.

By 1989 it had become unfashionable to discard fuel drums used in the course of field work, so next day Denny laboriously rolled them across to our side of the moraine. While feeling good that he was cleaning up the environment, he had an additional motive for moving them. They would make good runway markers.

Mark, Valerie, and I spent a bitterly cold day surveying the whole long line of stakes. Mark held the thirteen-foot measuring staff upright so that I could see it from afar through the levelling telescope, while Valerie noted down numbers as fast as I barked them out. Leap-frogging past Mark, Valerie drove the snowmobile while I suffered near-concussion in the sled while cushioning the instrument from shocks. After sighting backwards to Mark, he then leap-frogged past us.

The results were astonishing. Here was a potential runway long enough and level enough to operate any aircraft — including the largest now flying. We were excited by the thought of one day seeing

Traveling economy class in the Twin Otter.

the first Boeing 747 on its final approach to "Mill runway 36." That day could be a long way off, but if there was a need, nothing but precedent could stand in the way.

Stirring the beef stew that night I asked Valerie how she had come to sign on for a season at the South Pole. It transpired that she was very much an outdoor girl, having walked the length of the Appalachian Trail alone with her dog Buddy. At 26 years of age, she had notched up 500 hours as a helicopter pilot with the Vermont Army National Guard. She had worked at a home for mentally handicapped people and was still in school. After a couple of days our 37-year age-difference seemed to evaporate — we became just colleagues working together on a project. I wondered where else in the world the concept of rank or seniority could disappear as fast as it does in Antarctica. It becomes irrelevant.

Temperatures were still a balmy +10° F but 7 January dawned with a wind so vicious that there was nothing for it but to stay in our sleeping bags and read books. Every day we heard LC-130 supply flights droning back and forth between McMurdo and the pole because we were exactly on their track. Having nothing better to do, we called them on the radio from time to time and sent messages that we had failed, with our low-power radio, to get through to the South Pole. Valerie, steeped in history, was reading Paul Siple's *90° South*,[1] the story of the first South Pole Station erected in 1956.

The next day, in spite of winds varying from 15 to 30 knots, we managed to set out and survey two stake lines perpendicular to the main runway — though I had to

Malcolm Mellor.

keep one hand on the instrument tripod to prevent it crashing to the ground. But my diary says "Everyone cheerful." By 9 January we had virtually finished what we had set out to do, so took time off to study the surroundings. Towards the middle of the glacier, the surface became rough and eventually crevassed, so there was no sense in planning any cross-glacier runway. Denny and I climbed 500 feet to a survey cairn on Plunket Point, finding on the way some boulders with bizarre shapes cut by eons of winds that have scoured this landscape.

Looking out over the confluence of the mighty Beardmore with Mill Glacier, its tributary, the scene was breathtaking. Before us, disappearing into the distance, lay the railroad-smooth lines of the medial moraine that we had crossed earlier. Beyond, the great bulk of The Cloudmaker marked the left bank of the glacier where it curves through a gentle bend. Ninety miles away, the sharp summit of Mount Kyffin showed where the ice reaches its final, tumbling icefall, before going afloat on the sea and merging with the Ross Ice Shelf. To our left stretched seemingly endless lines of crevasses and the great icefall that Shackleton and Scott had to struggle up to reach the relative calm of the polar plateau. We had maps to show the lie of the land — they had nothing but courage and some inner compulsion to drive themselves on towards the South Pole.

While Denny and I were admiring the view, Mark and Valerie had taken another long hike, this time up the right bank of the Beardmore and back across Meyer Desert. By midnight I was becoming concerned for their safety, and scanned the distant hills with binoculars for some sign of movement. It was not until an hour later, after they had been walking for 5½ hours, that I saw two weary specks moving towards the tents. They were so elated at having traversed a vast stretch of virgin territory that I did not have the heart to tell them of my anxious wait. I remembered that once, in 1960, I had berated Tom Taylor for staying out hours after he had led me to expect his return. But I lived to regret it. Really good field men and women are rare — and best handled with kid gloves.

Having been summoned, our Twin Otter came to pick us up on 10 January. The pilots were having a long day. First they flew to Mount Howe. Schutt and Harvey had found some meteorite fragments but now were glad to be taken back to Bill Cassidy's camp. After refuelling at the Snakeskin cache, we carried aboard — with great difficulty — both of the heavy rubber sealdrums that were now empty. It was a labor of love that saved USAP from having to replace these very costly containers. McMurdo had written them off as expendable. A few minutes later our party was jammed into the already full cabin and we set course for the pole.

Our work was now done. The aircraft and Denny Hartog flew back to McMurdo, but I stayed at the South Pole to await Malcolm Mellor who was coming to see the two good ice runways for himself.

Non-government visitors had never been welcome at the South Pole and only a few dozen had ever reached it. An unprecedented event on 17 January was the arrival of a party of seven skiers supported by four guides. Three of the guides — Martyn Williams (Canadian), Mike Sharp (British), and Alejo Contreras (Chilean), were old friends. The others were:

Colonel J. K. Bayad (45), Indian Army
Jerry Corr (56), realtor, USA.
Stuart Hamilton, Canada
Shirley Metz (37), businesswoman, USA.
Ronald Milmarik (45), USAF dentist
Victoria Murden (24), divinity student, USA.
Joseph Murphy (55), writer, USA.
James Williams, USA.

The party had skied all the way from Hercules Inlet on Ronne Ice Shelf, a distance of 771 miles, at an average speed of 15.7 miles per day. The colonel was the first of his countrymen to reach the South

Skiers arriving at the South Pole, 17 January 1989. They had skied 771 miles from Hercules Inlet via Patriot Hills in 49 days.

Pole overland, and the women also were the first overland. Each of them had spent nearly $100,000 to join this — their trip of a lifetime. To most people, the idea of spending big money in exchange for seven weeks of gruelling toil in below-zero temperatures, much of the time facing into icy headwinds, without once washing or changing clothes, is nearly incomprehensible. Yet — for some — such is the lure of the Antarctic wilderness.

The expedition had been organized by ANI, the Canadian company whose DC-4 aircraft had flown me from Punta Arenas to Patriot Hills the year before; and Mountain Travel, a California travel company. After ANI's five years of carrying paying passengers — as Antarctica's first and only commercial airline — this expedition was the apex of ANI's achievement.

The party's arrival at the South Pole was preceded, an hour earlier, by the landing of one of ANI's Twin Otters that had come to pick them up. Everyone on the station turned out to cheer the skiers as they approached in line abreast across the last half mile to the South Pole marker. They were invited inside for a conducted tour of the station. USAP policy places strict limits on the hospitality that may be offered to non-government visitors. Though the skiers' reception was genuinely friendly, they were given coffee and doughnuts — nothing more.

The Twin Otter pilot was Henry Perk,

A picnic by the hovercraft, with Mount Erebus behind.

a Swiss-Canadian who had brought along his father as copilot. He had carried four drums of jet fuel for the return flight to Thiel Mountains. This range of mountains at latitude 85° S was the site of a half-way fuel cache between the pole (90° S) and Patriot Hills (80° S). The tenuous logistic lifeline needed to carry fuel over the 2,600 airline miles from Punta Arenas was underscored by the company's estimate that each 50-gallon drum was worth $24,000 by the time it reached the South Pole. No other airline on Earth operates over such extreme distances while retaining enough fuel to ferry each of its aircraft back to South America at the end of the season.

Before leaving, Henry Perk handed me a glove that I had lost at Patriot Hills 12 months earlier. Hoping to see me some day, he had kept it for a year. It reminded me of something learned over the years, that regardless of nationality, the Antarctic community is a fellowship of friends.

Malcolm Mellor arrived in McMurdo's second Twin Otter (also Canadian-registered) on 21 January. After exchanging greetings, we gravitated towards a wild Pisco party in the South Pole bar. Pisco is the native fire-water of Chile, and Henry Perk had presented a case of it to the South Pole Station when he came to collect the skiers. The custom of bringing a gift for one's host prevails throughout the Antarctic even to the highest latitudes, in spite of the exorbitant cost — in this case — of flying it all the way from Chile.

The next morning, when our heads had cleared, we flew Malcolm to the two prospective ice runways, first to Mill Glacier and then to Mount Howe. Malcolm was a scientist of extraordinary versatility. His research had spanned nearly every branch of pure and applied glaciology, from the mass balance of ice sheets to cold regions engineering, avalanches, drifting snow, icebergs, sea-ice, and permafrost. It was a delight to discover that he was as enthusiastic as I about what we had found.

I left for home shortly afterwards. In the course of 72 flying hours we had found two potential ice runways suitable for operating even the heaviest wheeled aircraft. We had made eight wheels-down landings at Mount Howe and five on Mill Glacier — all without any difficulty. Through visiting many other blue icefields we had ruled out the possibility of finding better sites within a reasonable distance from the South Pole. Cargo aircraft weighing up to 400 ton fly elsewhere — so why not on ice in Antarctica? Anything that could be done to make Antarctic logistics more cost-effective would make the research enterprise itself more cost-effective. Large aircraft use less fuel (per ton/mile) than smaller aircraft and reduce atmospheric pollution when compared with moving the same tonnage with smaller aircraft.

Our meteorite, a serendipitous by-product of the season, turned out to be more valuable than expected. Though iron meteorites are not uncommon, this one was unique. Its chemical composition aroused great interest at the Smithsonian Institution and became the subject of a separate research paper at a meteorite conference in Australia.[2] Among other minerals, it contained schreibersite, kamacite, and taenite, none of which I had ever heard of.

I reported on our ice runway findings[3] while Malcolm set about trying to convince the National Science Foundation and their Navy pilots that the acid test would be to land an LC-130 wheels-down on Mill Glacier. Some found the idea "revolutionary," others "crazy." In between, there were skeptics. We kept up the pressure by preparing — jointly — a detailed analysis of all aspects from an engineering point of view.[4]

The following year, with Malcolm Mellor in the cockpit, an LC-130 made a series of low passes over Mill Glacier on 26 January 1990 before touching down close to where our camp site had been the year before. I had advised landing a Twin Otter ahead of the Hercules to mark a runway and to serve as a weather-station. They had done neither. In one step, the aviators had traded overcaution for incaution. As the crew stepped onto the ice, one of them remarked "What's the problem?"

There was no problem.

---

1 Paul Siple. *90° South. The Story of the American South Pole Conquest.* New York, G. P. Putnam's Sons, 1959.

2 R. S. Clarke, Jr., V. F. Buchwald, and E. Olson. Anomalous ataxite from Mount Howe, Antarctica. *Meteoritics*, Vol. 25, No. 354, 1990, p. 354.

3 Charles Swithinbank. *Ice runways near the South Pole.* Hanover, New Hampshire, US Army Cold Regions Research and Engineering Laboratory (CRREL Special Report 89-19), 1989.

4 Malcolm Mellor and Charles Swithinbank. *Airfields on Antarctic glacier ice.* Hanover, New Hampshire, US Army Cold Regions Research and Engineering Laboratory (CRREL Special Report 89-21), 1989.

Epilogue

> *Be thou as chaste as ice, as pure as snow, thou shalt not escape calumny.*
>
> William Shakespeare (1564–1616)

I look back on my adventures in Antarctica with immense satisfaction. In six seasons of field work, my colleagues and I accomplished most of what we set out to do. The 1959–1960 crossing from Kainan Bay to McMurdo was repeated with tellurometers in subsequent seasons, leading to better measurements of the velocity of the ice shelf.[1] The valley glacier work determined the rate of movement of seven glaciers including some of the largest in the world.[2] Combined with gravity measurements, the work gave a total volume discharge for the seven glaciers studied of about eight cubic miles per year.[3] Later airborne radio-echo sounding of the ice thickness changed some of the figures for individual glaciers but made only a small difference to the total.[4]

Our pioneering work in seeking to understand the dynamics of the Ross Ice Shelf turned out to have an unexpected relevance to today's world. In 1959, it was seen, even by some glaciologists, as interesting but esoteric. Yet since 1963, USAP has spent well over a hundred million dollars elaborating on it. Why? The answer lies in what is now called global change. The world is changing and with it the ice. When, long ago, my colleagues and I postulated that the Antarctic ice sheet served as the principal control on world sea level, we were voices in the wilderness. Now this is common knowledge. In recent years, some ice shelves have virtually disappeared.[5] Others are evidently in the process of breaking up.[6] Although ice shelf fluctuations do not themselves change sea level, ice shelves serve to stabilize parts of the vast inland ice sheet resting on rock. There are indications that the volume of ice on land is changing — sea level changes must follow.

Globally-averaged mean temperatures have shown a net warming over the last 130 years during which sea level has risen.[7] In the Antarctic Peninsula area there has been an observed climatic warming of 2.7° C over the past half-century.[8] Though there are compensating mechanisms (negative feedbacks) in the climate system, we can only expect further changes. Glaciologists are being asked to predict what will happen during the 21st century, but we still have insufficient understanding of the interactions involved in climatic change. Research goes on. With it comes a growing realization that mankind is now inadvertently changing the face of the Earth.[9] Changes in the amount of ice will prove a crucial part of the story.

Throughout this account I have tried to strike a fair balance between praise and criticism of the vast US investment in Antarctica. USAP is the most productive research enterprise in the history of the continent and I have always known what a privilege it was to take part in it. There is no paradox in adding that the largest research enterprise is not the most cost-effective. US Government policy has, until quite recently, required the use of military personnel to support the program.[10] One of the design criteria for a fighting force is that each unit must be able to function effectively after decimation by an enemy. It follows that, in the absence of enemies, military organizations find it hard to come to terms with their own over-manning.

Over the past 20 years, USAP has made great strides in reducing the military presence within its operations, yet there remain major parts that could be handed to civilian contractors at a saving to the taxpayer. Reports from Washington speak of " . . . the Navy's eagerness to get out of the business of providing logistic support for science — a mission that now falls outside its military responsibilities."[11] A number of countries working in Antarctica manage not only their science but also their logistic support with no, or almost no, military involvement. Among them are Australia, China, France, Germany, Poland, Russia, and Britain. While none of these can match the US in its total contribution to science, all make more cost-effective use of their much smaller budgets.

In saying these things I may be accused of biting the hand that fed me. That is not the way I see it. On many occasions and often in difficult conditions, US military personnel have helped me far above and beyond the call of duty. They have been unfailingly cheerful, easy and stimulating to work with. Some have become friends for life. I would never advocate changing the type of people involved — military or civilian. All I would like is to see them welded into a force whose loyalty was to USAP as a whole rather than to its component parts. It could be done.

To the regret of many of them, Navy personnel are often rotated, taking them out of Antarctic work just when they are becoming adjusted to their job and to the environment. The result is a loss of vital accumulated experience. People learn best from their own experiences. We may commit things to paper for the benefit of the next generation but we cannot ensure that they are read.

One example: In January 1960, as my party crossed a crevassed area near McMurdo that had been known about for almost 60 years, I saw the upended runners of a 20-ton sledge that had followed its 35-ton tractor into a crevasse, fortunately without loss of life. The accident was well-publicized at the time — clearly this was no place for heavy vehicles. Thirty years later a similar vehicle towing a similar sledge fell into a crevasse at the same place.[12] Loss of accumulated experience is far from being unique to our calling but I still wonder how to hand on the torch.

What has become of my retirement project of reducing the cost of access to the interior by using wheeled aircraft rather than ski aircraft? The Canadian company Adventure Network International, which led the way in 1987 with an unmodified Douglas DC-4, replaced it with a pressurized DC-6B in 1989 and by a Lockheed L-382G Hercules in 1993, allowing faster round trips and greater payloads.[13] USAP has several times landed their LC-130 aircraft on the blue ice runway at Mill Glacier and at Patriot Hills. They have considered the use of wheeled aircraft to bring in materials for a new South Pole Station.[14] Giles Kershaw intended to use jet aircraft in due course and there is no doubt that someone will do it. Sadly, Giles did not live to see the day; he died in a flying accident in March 1990.[15] Malcolm Mellor also went the way of all flesh; he died at his home in 1991 at the age of 58.[16]

In olden days, explorers considered it their prerogative to name geographical features after their sponsors — or spouses. But today, few major features remain unnamed, so the process of naming is strictly controlled. The US Board on Geographic Names awards names at its discretion. Of colleagues I have worked closely with since 1959, it gladdens my heart to see on official maps the names:

> Cameron Island
> Cape Hickey
> Crary Mountains
> Evans Ice Stream
> Hughes Ice Piedmont
> Linder Peak
> Long Gables
> Mellor Glacier
> Mount Brecher
> Mount Krebs
> Mount Tuck
> Olson Nunatak
> Robin Peak
> Rundle Peaks
> Schroeder Peak
> Swithinbank Range
> Taylor Nunatak
> Thiel Mountains
> Zumberge Coast

As the years pass, so do some of my colleagues. While feeling keenly their loss, I look back with satisfaction on what we did together. It is remarkable how many of the great initiatives in Antarctic exploration have been the work, not of governments, nor committees, nor faceless bureaucrats, but instead of individuals who had ideas and pursued them with their heart and soul. Jim Zumberge was certainly one of them. After a distinguished career both as scientist and university president, he died in 1992 at the age of 68.[17] Jack Tuck died in 1984;[18] Bert Crary in 1987.[19]

Science, however, looks forward — not backward. The achievements of a lifetime are seen only in the context of contemporary knowledge. Today, most of the work that we did could be completed in a fraction of the time. Should we have waited for the means available today?

That question can be posed only with hindsight, so the answer can only be "No." At all stages of our work we have used the best means available — we could not know what developments lay ahead. Marco Polo used horses; Christopher Columbus used sailing ships; Amundsen drove dogs. Nothing better was available.

We drove snowmobiles because there were no dogs. Instruments have improved, living conditions have improved, aircraft have improved. People remain the same — ever striving to improve their lot.

Traveling within the Antarctic has become safer because more use is made of aircraft. Admiral Byrd was the great pioneer in the wider use of aircraft. In relation to the perceived hazards, Antarctic aviation has on the whole a good safety record. Although we learn to cope with white-out, blizzards, ice in fuel, and sometimes only sketch maps to navigate with, at the same time we are blessed with the world's biggest potential landing ground for ski-equipped aircraft — five million square miles of it. On New Year's Day 1962, a Navy LC-130 flying over the Siple Coast lost power on three of its four engines owing to ice in the fuel lines; the pilot was forced to land where he was. The plane made a perfect landing, and after clearing the ice, flew back to McMurdo.[20] Where else in the world could there have been such a happy outcome? At home the pilot in a similar emergency would be fighting to avoid buildings, earthworks, trucks, trees, cattle, and a host of other obstructions.

One of the great changes that I have witnessed has been in the type of people interested in the Antarctic. Forty five years ago the Norwegian-British-Swedish expedition, of which I was a member, attracted interest from newspapers in many countries, but the interest was ephemeral. The Antarctic was soon forgotten. The great advantage of being forgotten — compared with now — was that we had the Antarctic to ourselves, without interference from governments, international lawyers, or environmentalists bent on curbing access to the continent.

Environmentalists with a gift for objectivity — most of them independent of governments — have done a great deal to draw attention to avoidable human impacts on the Antarctic environment. Many research stations are today much tidier and their occupants are more careful about the disposal of rubbish than they would have been if Greenpeace had not persisted in drawing attention to the long-term effects of neglect. However, by promulgating myths, or being economical with the truth, some environmental organizations — including Greenpeace — have alienated their most effective potential partners and allies — the scientific establishment.

Today, we are besieged by a conspiracy of misinformation about the motives of science. The public is told that government support for Antarctic science has been covertly dedicated to winning the race to exploit the vast mineral wealth of Antarctica. A magazine headline reads: Once inaccessible and pristine, the white continent is now threatened by spreading pollution, budding tourism, and the world's thirst for oil.[21] People are not told that the race is but an artifact of minds geared to material gain. They are not told that the average thickness of ice on the continent is eight thousand feet.[22] The truth is that mining or extracting oil through great depths of moving ice is now and will long remain beyond the dreams of avarice.

They are not told that ice-free land represents — not three or four percent of the area of the continent as reported in many published accounts — but just 0.33

percent.[23] This is smaller than the state of West Virginia. But unlike West Virginia, no mineral resources have been found in it, and if they ever are, they are hardly likely to be worth exploiting.

Antarctica is still the only continent that we can visit without a passport. The Antarctic Treaty has been a phenomenal success story, keeping the peace throughout the Cold War and in spite of conflicting — and still unresolved — claims to sovereignty. Signed by 12 states in 1959, the ensuing years have seen accessions by a further 30 states.[24] Taken together, the signatories now represent about 80 percent of the population of the globe.

There have been strains but no real quarrels. For years, newspapers carried the story that the Antarctic Treaty was to expire in 1991, therefore had to be renewed or renegotiated with urgency before anarchy ensued. It seems to have been accepted as fact by 90 percent of the population. Even scientists spread the story, perhaps because of their own hidden agenda. Evidently, few of them had read the terms of the actual treaty, 14 brief articles that can be read in 10 minutes. No words such as renewal or renegotiation are used anywhere in the treaty. It has no time limit. The preamble, which sets the tone of the document, explicitly recognizes:

> . . . that it is in the interest of all mankind that Antarctica shall continue forever to be used exclusively for peaceful purposes and shall not become the scene or object of international discord . . .

Another contemporary problem is that people are led to believe that the continent is overrun by scientists and others who do not care about the environment. Yet the total wintering population is less than 1,200.[25] Antarctica, far from being overrun, is in winter about 25,000 times less densely populated than Australia. A continent larger than the United States holds the population of a small village. Adding short-term visitors, perhaps 10,000 summer support personnel[26] and 10,000 tourists (both figures are greater than any yet reported), we could — each one of us — have 250 square miles to ourselves.

There are, however, concentrations of population. For example, on King George Island in the South Shetland Islands, there are year-round research stations belonging to eight countries: Argentina, Brazil, Chile, China, Korea, Poland, Russia, and Uruguay. Other countries have a summer-only presence. Taken together, these must have a substantial impact on the environment — but only locally. Concentrations of population leave greater areas of Antarctica untouched. Perhaps we should encourage them.

A former President of the international Scientific Committee on Antarctic Research (SCAR), himself a highly respected scientist, maintains that 99.999 percent of the area of Antarctica "remains virtually unaffected by the impact of human activity."[27] On the initiative of scientists, 35 places in Antarctica have been designated "Sites of Special Scientific Interest" (SSSI) and 19 areas have been designated "Specially Protected Areas" (SPA). All of these are regarded as Conservation Areas and subject to strict rules of entry. Entry to SPAs is prohibited except with a permit issued by an Antarctic Treaty government.

SCAR also provided scientific advice

to guide those who drafted the 1991 *Protocol on Environmental Protection to the Antarctic Treaty*.[28] This contains stringent and detailed regulations on measures to minimize human impact on the environment. Article Seven prohibits mining and even prospecting. The agreement covers many matters that were omitted from the 1959 Antarctic Treaty because of possible conflicts of interest between states. Thus international control of activities in Antarctica is becoming stronger than ever.

The provisions of the Protocol apply equally to government and non-government visitors to the Antarctic. Non-government visitors, including scientists, mountaineers, skiers, tourists, and adventurers, have been a target for ill-informed criticism on environmental grounds. Hordes of tourists are said to be trampling the vegetation, upsetting the wildlife, littering the continent, and polluting the sea.

It would be hard to imagine anything further from the truth. The only scientist who has attempted to quantify the effects of tourism on the Antarctic environment concluded that tourism in all its forms represents about 0.5 percent of the total human impact.[29]

Greater numbers of visitors will increase the impact and we must seek a balance between access and excess. Certainly there must be monitoring and control — and there will be under the terms of the Protocol.

How is it that people who believe that they are well informed can entertain such divergent views? I do not question the motives either of environmentalists or of scientists — they have a common interest in preserving Antarctica in its pristine state. I only deplore the alarmists.

Luckily for both sides of the argument, most activities in Antarctica are self-limiting and will remain so. There are no natural resources that can sustain an indigenous population. The number of scientists is limited by the very high cost of importing everything necessary for their work and survival. Scientists with no interest in the polar regions are vociferous in pointing out how much better off their own research might be if less was spent on Antarctic work. It is no secret that dozens of scientists could be employed back home for the cost of a single one in Antarctica.

Similarly, sport in Antarctica is not for everyman. Mountaineers pay $26,000 each to climb Vinson Massif and skiers have paid more than $70,000 for the ordeal of skiing to the South Pole. Costs will come down but can only remain high compared with similar activities on any other continent.

Finally, who are the explorers today? The term has come to have a broader meaning than ever in the past. It used to be kept for those who had trudged over the surface to map unknown lands. Later it was held to include airmen who had merely flown over new land and photographed it. Modern scientists — though some have indeed explored in the classic sense — have rejected the term in order to distinguish themselves from mere travelers. Today the definition seems to have widened to include travelers who, to test their physical endurance, use obsolete methods of travel like man-hauling.

Astronauts explore space, astronomers the universe? Physicists explore the structure of matter. Biologists explore genes. Psychologists explore the mind.

> In my view, in one sense or another, we are all explorers.

1 Egon Dorrer, Walther Hofmann, and Wilfried Seufert. Geodetic results of the Ross Ice Shelf Survey Expeditions, 1962–63 and 1965–66. *Journal of Glaciology*, Vol. 8, No. 52, 1969, pp. 67–90.

2 Charles W. Swithinbank. Ice movement of valley glaciers flowing into the Ross Ice Shelf, Antarctica. *Science*, Vol. 141, No. 3580, 1963, pp. 523–524.

3 M. Giovinetto, Edwin S. Robinson, and C. W. M. Swithinbank. The regime of the western part of the Ross Ice Shelf drainage system. *Journal of Glaciology*, Vol. 6, No. 43, 1966, pp. 55–68.

4 Charles Swithinbank. *Satellite image atlas of glaciers of the world: Antarctica*. US Geological Survey Professional Paper 1386B, 1988.

5 C. S. M. Doake and D. G. Vaughan. Rapid disintegration of the Wordie Ice Shelf in response to atmospheric warming. *Nature*, Vol. 350, No. 6316, 1991, pp. 328–330.

6 Pedro Skvarca. Changes and surface features of the Larsen Ice Shelf, Antarctica, derived from Landsat and Kosmos mosaics. *Annals of Glaciology*, Vol. 20, 1994, pp. 6–12.

7 J. T. Houghton, L. G. Meiro Filho, B. A. Callander, N. Harris, A. Kattenburg and K. Maskell(eds.). *Climate Change 1995. The science of climatic change. Contribution of WGI to the second assessment report of the Intergovernmental Panel on Climate Change.* Cambridge, Cambridge University Press, 1996.

8 P. Stark. Climatic warming in the central Antarctic Peninsula area. *Weather*, Vol. 49, No. 6, 1994, pp. 215–220.

9 Richard A. Kerr. Studies say — tentatively — that greenhouse warming is here. *Science*, Vol. 268, No. 5217, 1995, pp. 1567–1568.

10 Ronald Reagan. Memorandum of 5 February 1982 addressed to the Secretary of State, the Director of the National Science Foundation, and others.

11 Jeffrey Mervis. NSF eyes new South Pole Station. *Science*, Vol. 264, No. 5167, 1994, p. 1838.

12 *Antarctic* (a news bulletin published quarterly by the New Zealand Antarctic Society), Vol. 12, No. 2/3, 1990, p. 54.

13 Charles Swithinbank. Airborne tourism in the Antarctic. *Polar Record*, Vol. 29, No. 169, 1993, pp. 103–110.

14 Stephen L. DenHartog and George L. Blaisdell. *Delivery of fuel and construction materials to South Pole Station.* Hanover, New Hampshire, US Army Cold Regions Research and Engineering Laboratory (CRREL Special Report 93-19), 1993.

15 Charles Swithinbank. Obituary: John Edward Giles Kershaw. *Polar Record*, Vol. 26, No. 158, 1990, p. 250.

16 Charles Swithinbank. Obituary: Malcolm Mellor. *Polar Record*, Vol. 28, No. 164, 1992, p. 80.

17 Charles Swithinbank. Obituary: James Herbert Zumberge. *Polar Record*, Vol. 28, No. 167, 1993, p. 337.

18 Obituary: John Tuck, Jr. *Polar Record*, Vol. 22, No. 139, 1985, p. 454.

19 Gordon de Q. Robin. Obituary: Albert P. Crary. *Polar Record*, Vol. 24, No. 149, 1988, pp. 147–148.

20 David M. Tyree. New era in the loneliest continent. *National Geographic Magazine,* Vol. 123, No. 2, 1963, p. 283.

21 *TIME*, 15 January 1990, p. 55.

22 D. J. Drewry, S. R. Jordan, and E. Jankowski. Measured properties of the Antarctic ice sheet: surface configuration, ice thickness, volume and bedrock characteristics. *Annals of Glaciology*, Vol. 3, 1982 (p. 83).

23 Adrian J. Fox and Paul R. Cooper. Measured properties of the Antarctic ice sheet derived from the SCAR Antarctic digital database. *Polar Record*, Vol. 30, No. 174, 1994, pp. 201–206.

24 The original 12 signatories of the Antarctic Treaty were (in order of their ratification): United Kingdom, South Africa, Belgium, Japan, United States, Norway, France, New Zealand, USSR, Argentina, Australia, and Chile. Accessions (in chronological order) were: Poland, Czechoslovakia, Denmark, Netherlands, Romania, East Germany (DDR), Brazil, Bulgaria, West Germany (BRD), Uruguay, Papua New Guinea, Italy, Peru, Spain, People's Republic of China, India, Hungary, Sweden, Finland, Cuba, South Korea, Greece, North Korea, Austria, Ecuador, Canada, Colombia, Switzerland, Guatemala, and Ukraine.

25 *The World Factbook 1995*. Washington, DC, Central Intelligence Agency, 1995.

26 Juan Carlos M. Beltramino. *The Structure and Dynamics of Antarctic Population.* New York, Vantage Press, 1993.

27 Richard Laws. Unacceptable threats to Antarctic science. *New Scientist*, 30 March 1991, p. 4.

28 Protocol on Environmental Protection to the Antarctic Treaty. *Polar Record*, Vol. 29, No. 170, 1993, pp. 256-275.

29 R. K. Headland. Historical development of Antarctic tourism. *Annals of Tourism Research*, Vol. 21, No. 2, 1994, pp. 269–280.

Acronyms

ANI	Adventure Network International
CRREL	Cold Regions Research and Engineering Laboratory of the United States Army Corps of Engineers.
FACH	Fuerza Aérea de Chile (Chilean air force).
JATO	Solid fuel rockets used to assist take-off; abbreviation for Jet Assisted Take Off.
NSF	National Science Foundation (Washington, DC)
NZAP	New Zealand Antarctic Programme
NZARP	New Zealand Antarctic Research Programme
SCAR	Scientific Committee on Antarctic Research (of the International Council of Scientific Unions).
USAP	United States Antarctic Program
USARP	United States Antarctic Research Program

Glossary

Altimeter	Aneroid barometer used to measure altitude of aircraft.
Banana sled	Fiberglass sled used for hauling light loads. Sometimes pulled with rigid (bamboo) traces.
Barrier	Obsolete name for ice front and/or ice shelf.
Cache	Supplies left in the field for later use.
Caldera	Large scale volcanic crater often formed by subsidence.
Calving	The breaking away of a mass of ice from a floating ice shelf, glacier, or iceberg.
Cornice	An overhanging accumulation of wind-blown snow on the edge of a crevasse.
Crampon	Spiked metal frame strapped to boots to facilitate climbing on ice.
Crevasse	A fissure formed in a glacier, ice sheet, or ice shelf. Crevasses are often hidden by snow bridges.
Cryoconite	Wind-transported rock debris or plant material on a glacier. Cryoconite holes are formed where it melts its way into the ice by absorbing solar radiation.
Depot or **depôt**	Supplies left in the field for later use. The British equivalent of the more common American cache.
Dolerite	Volcanic rock similar to basalt, containing feldspar and augite.
Dunnage	Wood used between layers of a ship's cargo to prevent chafing.
Fast ice	Sea-ice which remains fast along the coast, where it is attached to the shore or to an ice shelf. An abbreviation of landfast ice.
Floe	A piece of floating ice other than fast ice or glacier ice.
Galley	Naval term for kitchen. In practice used to include dining room.
Hypothermia	Cooling of the body to danger level as a result of heat loss from exposure.

Ice drill	A coring device for obtaining ice samples. May be hand- or motor-powered.
Icefall	A heavily crevassed area in a glacier at a region of steep descent.
Ice front	The vertical cliff forming the seaward face of an ice shelf or other floating glacier, varying in height from six to 150 feet above sea level.
Ice piedmont	Ice covering a coastal strip of low-lying land backed by mountains.
Ice sheet	A mass of ice and snow of considerable thickness and large area. Ice sheets may be resting on rock (see Inland ice sheet) or floating (see Ice shelf). Ice sheets of less than 50,000 square kilometers in area are called ice caps.
Ice shelf	A floating ice sheet of considerable thickness attached to a coast. Ice shelves are usually of great horizontal extent and have a level or gently undulating surface. They are nourished by the accumulation of snow and often by the seaward extension of land glaciers. Limited areas may be aground. The seaward edge is termed an ice front.
Ice stream	Part of an ice sheet in which the ice flows more rapidly and not necessarily in the same direction as the surrounding ice. The margins are sometimes clearly marked by a change in direction of the surface slope, but may be indistinct.
Inland ice sheet	An ice sheet of considerable thickness and more than about 50,000 square kilometers in area, resting on rock. Inland ice sheets near sea level may merge into ice shelves.
Intersection	Determining the position of a distant point by measuring angles from each end of a line of known length.
Katabatic wind	Air cooled over higher ground flowing to lower levels.
Knot	A unit of speed equal to one nautical mile per hour.
Krill	A small crustacean *Euphausia superba*.
Mirage	Optical phenomenon in which distant objects appear uplifted above the horizon. Caused by abnormal refraction with a surface temperature inversion in which air temperature increases with height.
Moraine	Ridges or deposits of rock debris transported by a glacier. Common forms are: lateral moraine, along the sides; medial moraine, down the center.

Névé	Old snow which has been transformed into a dense material.
Nunatak	A rocky crag or small mountain projecting from and surrounded by a glacier or ice sheet.
Pack ice	An area of sea-ice other than fast ice, no matter what form it takes or how it is disposed.
Plane table	A drawing board mounted horizontally on a tripod. A sight rule is used to record directions to objects for mapping.
Resection	Determining an observer's position by measured angles to points whose coordinates are known. The converse of intersection.
Sastrugi	Sharp, irregular ridges formed on a snow surface by wind erosion and deposition. The ridges are usually parallel to the prevailing wind.
Sea-ice	Any form of ice found at sea which originated from the freezing of sea water.
Sérac	A sharp ridge or pinnacle of ice on a glacier or icefall.
Skidoo	An alternative word for snowmobile.
Sno-cat	An oversnow vehicle sprung over four separate tracks.
Snow bridge	An arch formed by snow which has drifted across a crevasse, forming first a cornice, and ultimately a covering which may completely obscure the opening.
Snowdrift	An accumulation of wind-blown snow deposited in the lee of obstructions or heaped by wind eddies.
Snowmobile	Single-tracked vehicle with steerable ski at the front.
Stadia rod	A thirteen foot high graduated staff used with an engineer's (telescopic) level to determine height differences between points.
Strand crack	A fissure at the junction between an inland ice sheet, ice piedmont or ice rise and an ice shelf, the latter being subject to the rise and fall of the tide.
Sublimation	Conversion of ice or snow to vapor without passing through the liquid phase.
Tellurometer	Electronic distance-measuring equipment. One instrument is set up at each end of the line to be measured.
Theodolite	A precise angle-measuring instrument consisting of a telescopic sight mounted on graduated horizontal and vertical circles.

Trilateration	Measuring the length of all sides of triangles to make a survey network.
Valley glacier	A glacier which flows down a valley.
Wanigan	A shelter mounted on a sledge towed behind a tractor.
Weasel	Amphibious tracked snow vehicle (weighing two ton).
Whiteout	A condition in which daylight is diffused by multiple reflection between a snow surface and overcast sky. Contrasts vanish and the observer is unable to distinguish snow surface features.

References

Amundsen, Roald. *The South Pole. An Account of the Norwegian Antarctic Expedition in the "Fram", 1910–1912*. New York, Lee Keedick, 1913 (2 vols).

Back, J. D. (ed.). *The Quiet Land, the Diaries of Frank Debenham*. Huntingdon, Bluntisham Books; and Harleston, Erskine Press, 1992.

Bailey, J. T., and S. Evans. Radio echo-sounding on the Brunt Ice Shelf and in Coats Land, 1965. *British Antarctic Survey Bulletin* No. 17, 1968, pp. 1–12.

Beltramino, Juan Carlos M. *The Structure and Dynamics of Antarctic population*. New York, Vantage Press, 1993.

Blackburn, Quin A. Handwritten note left in cairn at Durham Point, 29 November 1934. Scott Polar Research Institute Archives (unpublished).

Blackburn, Quin A. The Thorne Glacier section of the Queen Maud Mountains. *Geographical Review*, Vol. 27, No.4, 1937 (p. 598).

Bogard, D. D., L. E. Nyquist, and P. Johnson. Noble gas contents of shergottites and implications for the Martian origin of SCN meteorites. *Geochim. Cosmochim. Acta*, Vol. 48, 1984, pp. 1723–1739.

Bogorodsky, V. V., C. R. Bentley, and P. E. Gudmandsen. *Radioglaciology*. Boston, D. Reidel Publishing Company, 1985.

Brecher, H. H. Surface velocity determination on large polar glaciers by aerial photogrammetry. *Annals of Glaciology*, Vol. 8, 1986, pp. 22–26.

Brunk, Karsten, and Rudolf Staiger. Nachmessungen an Pegeln auf einem Blaueisfeld im Borgmassiv, Neuschwabenland, Antarktis. *Polarforschung*, Vol. 56, No. 1/2, 1986, pp. 23–32.

Burke, David. *Moments of Terror. The Story of Antarctic Aviation*. Kensington, New South Wales Press, 1994 (p. 273).

Byrd, Richard Evelyn. *Little America. Aerial Exploration in the Antarctic, the Flight to the South Pole*. New York, G. P. Putnam's Sons, 1930.

Byrd, Richard Evelyn. *Discovery. The Story of the Second Byrd Antarctic Expedition*. New York, G. P. Putnam's Sons, 1935.

Byrd, Richard E. *Alone*. London, Putnam, 1938.

Byrd, Richard Evelyn. Our navy explores Antarctica. *National Geographic Magazine*, Vol. 92, No. 4, 1947, pp. 429–522.

Cassidy, William A. Meteorite search at Lewis Cliff ice tongue: Systematic recovery program completed. *Antarctic Journal of the United States*, Vol. 24, No. 5, 1989, p. 44.

Central Intelligence Agency. *The World Factbook 1995*. Washington DC., 1995.

Cherry-Garrard, Apsley. *The Worst Journey in the World*. London, Chatto and Windus, 1922 (2 vols).

Chipman, Elizabeth. *Women on the Ice. A History of Women in the Far South*. Melbourne University Press, 1986.

Clarke, R. S. Jr., V. F. Buchwald, and E. Olson. Anomalous ataxite from Mount Howe, Antarctica. *Meteoritics*, Vol. 25, No. 354, 1990, p. 354.

Crary, A. P. Letter to Glenn E. Bowie, 16 October 1963 (unpublished).

Darlington, Jenny. *My Antarctic Honeymoon: a Year at the Bottom of the World*. Garden City, N. Y., Doubleday, 1956.

Debenham, F. A new mode of transportation by ice. *Quarterly Journal of the Geological Society*, London, Vol. 75, Part 2, 1920, pp. 51–76.

DenHartog, Stephen L., and George L. Blaisdell. *Delivery of fuel and construction materials to South Pole Station*. Hanover, New Hampshire, US Army Cold Regions Research and Engineering Laboratory (CRREL Special Report 93–19), 1993.

Dewart, Gilbert. *Antarctic Comrades*. Columbus, The Ohio State University Press, 1989.

Doake, C. S. M., and D. G. Vaughan. Rapid disintegration of the Wordie Ice Shelf in response to atmospheric warming. *Nature*, Vol. 350, No. 6316, 1991, pp. 328–330.

Dorrer, Egon, Walther Hofmann, and Wilfried Seufert. Geodetic results of the Ross Ice Shelf Survey Expeditions, 1962–63 and 1965–66. *Journal of Glaciology*, Vol. 8, No. 52, 1969, pp. 67–90.

Drewry, D. J., S. R. Jordan, and E. Jankowski. Measured properties of the

Antarctic ice sheet: surface configuration, ice thickness, volume and bedrock characteristics. *Annals of Glaciology*, Vol. 3, 1982 (p. 83).

Eugster, O. History of meteorites from the Moon collected in Antarctica. *Science*, Vol. 245, No. 4923, 1989, pp. 1197–1202.

Evans, S., and G de Q. Robin. Glacier depth sounding from the air. *Nature* (London), Vol. 210, No. 5039, 1966, pp. 883–885.

Fiennes, Ranulph. *To the Ends of the Earth*. London. Hodder and Stoughton, 1983.

Fox, Adrian J., and Paul R. Cooper. Measured properties of the Antarctic ice sheet derived from the SCAR Antarctic digital database. *Polar Record*, Vol. 30, No. 174, 1994, pp. 201–206.

Fuchs, Sir Vivian, and Sir Edmund Hillary. *The Crossing of Antarctica*. London, Cassell, 1958.

Giaever, John. *The White Desert. The Official Account of the Norwegian-British-Swedish Antarctic Expedition*. New York, E. P. Dutton & Co., 1955.

Giovinetto Mario B., Edwin S. Robinson, and Charles W. M. Swithinbank. On the regime of the western part of the Ross Ice Shelf drainage system. *Journal of Glaciology*, Vol. 6, No. 43, 1966, pp. 55–68.

Gould, L. M. Handwritten note left in Amundsen's cairn at Mount Betty, 25 December 1929. Scott Polar Research Institute Archives (unpublished).

Gould, L. M. Handwritten note left in Amundsen's cairn at Mount Betty, 28 December 1929. Scott Polar Research Institute Archives (unpublished).

Gould, Laurence McKinley. *Cold, the Record of an Antarctic Sledge Journey*. New York, Brewer, Warren & Putnam, 1931.

Headland, R. K. Historical development of Antarctic tourism. *Annals of Tourism Research*, Vol. 21, No. 2, 1994, pp. 269–280.

Herbert, Wally. *A World of Men*. London, Eyre and Spottiswoode, 1968.

Herbert, Wally. *The Noose of Laurels*. London, Hodder and Stoughton, 1989.

Hillary, Sir Edmund. *No Latitude for Error*. London, Hodder and Stoughton, 1961.

Hillary, Edmund. *Nothing Venture, Nothing Win.* London, Hodder & Stoughton, 1975.

Houghton, John. T., L. G. Meiro Filho, B. A. Callander, N. Harris, A. Kattenburg, and K. Maskell (eds.). *Climate Change 1995. The Science of Climatic Change. Contribution of WGI to the second assessment report of the Intergovernmental Panel on Climate Change.* Cambridge, Cambridge University Press, 1996.

Hunt, John. *The Ascent of Everest.* London, Hodder & Stoughton, 1953.

Huxley, Leonard (ed.). *Scott's Last Expedition.* London, Smith Elder & Co., 1913, Vol. 1, Appendix (p. 631).

Joyce, Ernest. *The South Polar Trail. The Log of the Imperial Trans-Antarctic Expedition.* London, Duckworth, 1929.

Kellogg, Thomas B., Davida E. Kellogg, and Minze Stuiver. Late Quaternary history of the southwestern Ross Sea: Evidence from debris bands on the McMurdo Ice Shelf, Antarctica. American Geophysical Union, *Antarctic Research Series*, Vol. 50, 1990, pp. 25–56.

Kerr, Richard A., Studies say — tentatively — that greenhouse warming is here. *Science*, Vol. 268, No. 5217, 1995, pp. 1567–1568.

Kovacs, A., and G. Abele. *Runway site survey, Pensacola Mountains, Antarctica.* Hanover, New Hampshire: US Army Cold Regions Research and Engineering Laboratory (CRREL Special Report 77-14), 1977.

Laws, Richard. Unacceptable threats to Antarctic science. *New Scientist*, 30 March 1991, p. 4.

Likely, Wadsworth. *New York Herald Tribune*, 12, 14, and 16 December 1960.

MacDonald, Edwin A. Our icebreakers are not good enough. *United States Naval Institute Proceedings*, Vol. 92, No. 756, 1966, pp. 59–69.

Macfarlane, Stuart. *The Erebus Papers.* Auckland, Avon Press Ltd, 1991.

Mahon, P. T. *Report of the Royal Commission to Inquire into the Crash on Mount Erebus, Antarctica of a DC-10 Aircraft Operated by Air New Zealand Limited.* Wellington, Government Printer, 1981.

Mellor, Malcolm, and Charles Swithinbank. *Airfields on Antarctic glacier ice.* Hanover, New Hampshire, US Army Cold Regions Research and Engineering Laboratory (CRREL Special Report 89-21), 1989.

Mervis, Jeffrey. NSF eyes new South Pole Station. *Science*, Vol. 264, No. 5167, 1994, p. 1838.

National Science Foundation. *Survival in Antarctica*. Washington DC, National Science Foundation (n. d.).

National Science Foundation. Evacuation flight to Halley Bay. *Antarctic Journal of the United States*, Vol. 3, No. 1, 1968, pp. 14-15.

New Zealand Antarctic Society. Women = worries. *Antarctic*, Vol. 4, No. 1, 1965, p. 14.

New Zealand Antarctic Society. No women! *Antarctic*, Vol. 4, No. 1, 1965, p. 16.

New Zealand Antarctic Society. Joint US/NZ survival training pays off. *Antarctic*, Vol. 12, No. 2/3, 1990, p. 54.

New Zealand, Ministry of Transport. Air New Zealand McDonnell-Douglas DC10-30 ZK-NZP Ross Island, Antarctica, 28 November 1979. Aircraft Accident Report No. 79–139. Wellington, 1981.

Otway, Peter. Tin dogs versus shaggy dogs. *Antarctic*, Vol. 3, No. 6, June 1963, pp. 232–235.

Reagan, Ronald. Memorandum of 5 February 1982 addressed to the Secretary of State, the Director of the National Science Foundation, and others.

Reece, Alan. Sledges of the Norwegian-British-Swedish Antarctic Expedition, 1949–52. *Polar Record*, Vol. 6, No. 46, 1953, pp. 775–787.

Richards, R. W. The Ross Sea Shore Party 1914–17. *Scott Polar Research Institute Special Publication*, No. 4. Cambridge, Scott Polar Research Institute, 1962.

Robin, G. de Q., C. W. M. Swithinbank, and B. M. E. Smith. Radio echo exploration of the Antarctic ice sheet. Gentbrugge, *International Association of Scientific Hydrology*, Publication No. 86, 1970, pp. 97–115.

Robin, Gordon de Q. Obituary: Albert P. Crary. *Polar Record*, Vol. 24, No. 149, 1988, pp. 147–148.

Ross, Sir James Clark. *A Voyage of Discovery and Research in the Southern and Antarctic Regions during the years 1839–43*. London, John Murray, 1847.

Schytt, Valter. Blue ice-fields, moraine features and glacier fluctuations. *Norwegian-British-Swedish Antarctic Expedition 1949–52, Scientific Results*, Vol. 4E. Oslo, Norsk Polarinstitutt, 1960.

Scott, R. F. The Voyage of the *Discovery* (2 vols). London, Smith Elder & Co., 1905 (Vol. 1, p. 199).

Scott Polar Research Institute. New Zealand activities in the Antarctic, 1959–60. *Polar Record*, Vol. 10, No. 66, 1960, p. 280.

Scott Polar Research Institute. Obituary: John Tuck, Jr. *Polar Record*, Vol. 22, No. 139, 1985, p. 454.

Scott Polar Research Institute. Protocol on Environmental Protection to the Antarctic Treaty. *Polar Record*, Vol. 29, No. 170, 1993, pp. 256–275.

Seaver, George. *Edward Wilson: Nature-lover*. London, John Murray, 1937 (2 vols).

Seaver, George. *'Birdie' Bowers of the Antarctic*. London, John Murray, 1938.

Shackleton, E. H. *The Heart of the Antarctic, being the Story of the British Antarctic Expedition 1907–1909*. London, William Heinemann, 1909 (2 vols).

Shackleton, E. H. *South. The Story of Shackleton's Last Expedition 1914–1917*. London, William Heinemann, 1919.

Shackleton, Edward. *Nansen the Explorer*. London, Witherby, 1959.

Silverstein, Samuel C. The American Antarctic Mountaineering Expedition. *Antarctic Journal of the United States*, Vol. 2, No. 2, 1967, pp. 48–50.

Siple, Paul. *A Boy Scout with Byrd*. New York, G. P. Putnam's Sons, 1931.

Siple, Paul. *90° South. The Story of the American South Pole Conquest*. New York, G. P. Putnam's Sons, 1959 (p. 138).

Skvarca, Pedro. Changes and surface features of the Larsen Ice Shelf, Antarctica, derived from Landsat and Kosmos mosaics. *Annals of Glaciology*, Vol. 20, 1994, pp. 6–12.

Stark, P. Climatic warming in the central Antarctic Peninsula area. *Weather*, Vol. 49, No. 6, 1994, pp. 215–220.

Sullivan, Walter. *The New York Times*, 1 October 1969.

Swithinbank, Charles. *Norwegian-British-Swedish Antarctic Expedition*

1949–52, Scientific Results, Vol. 3. Oslo, Norsk Polarinstitutt, 1957–1960, pp. 1–159.

Swithinbank, Charles W. M., David G. Darby, and Donald E. Wohlschlag. Faunal remains on an Antarctic ice shelf. *Science*, Vol. 133, No. 3455, 1961, pp. 764–766.

Swithinbank, Charles. Motor sledges in the Antarctic. *Polar Record*, Vol. 11, No. 72, 1962, pp. 265–269.

Swithinbank, Charles. Ice movement of valley glaciers flowing into the Ross Ice Shelf, Antarctica. *Science*, Vol. 141, No. 3580, 1963, pp. 523–524.

Swithinbank, Charles. To the valley glaciers that feed the Ross Ice Shelf. *Geographical Journal*, Vol. 130, Part 1, 1964, pp. 32–48.

Swithinbank, Charles. Radio echo sounding of Antarctic glaciers from light aircraft. *International Association of Scientific Hydrology*, Publication No. 79, 1968, pp. 405–414.

Swithinbank, Charles. Antarctic Airways: Antarctica's first commercial airline. *Polar Record*, Vol. 24, No. 151, 1988, pp. 313-316.

Swithinbank, Charles. Satellite image atlas of glaciers of the world: Antarctica. *US Geological Survey Professional Paper* 1386B, 1988.

Swithinbank, Charles. *Ice runways near the South Pole*. Hanover, New Hampshire, US Army Cold Regions Research and Engineering Laboratory (CRREL Special Report 89-19), 1989.

Swithinbank, Charles. Obituary: John Edward Giles Kershaw. *Polar Record*, Vol. 26, No. 158, 1990, p. 250.

Swithinbank, Charles. Obituary: Malcolm Mellor. *Polar Record*, Vol. 28, No. 164, 1992, p. 80.

Swithinbank, Charles. Obituary: James Herbert Zumberge. *Polar Record*, Vol. 28, No. 167, 1992, p. 337.

Swithinbank, Charles. Airborne tourism in the Antarctic. *Polar Record*, Vol. 29, No. 169, 1993, pp. 103–110.

TIME, 15 January 1990, p. 55.

Tyree, David M. New era in the loneliest continent. *National Geographic Magazine*, Vol. 123, No. 2, 1963, p. 283.

US Antarctic Service. Reports of the scientific results of the United States Antarctic Service Expedition 1939–41. *Proceedings of the American*

Philosophical Society, Vol. 89, No. 1, 1945, pp. 1–398.

Vette, Gordon, with John Macdonald. *Impact Erebus*. Auckland, Hodder and Stoughton, 1983.

Victor, Paul-Émile. *Mes aventures polaires*. Paris, Éditions G. P., 1975.

Wilson, Charles R., and A. P. Crary. Ice movement studies on the Skelton Glacier. *Journal of Glaciology*, Vol. 3, No. 29, 1961, pp. 873–878.

Zotikov, I. A., and A. J. Gow. The thermal and compositional structure of the Koettlitz Ice Tongue, McMurdo Sound, Antarctica. In: International Conference on Low temperature Science, Sapporo, Japan, April 14–19, 1966, *Physics of Snow and Ice: Proceedings*, Vol. 1, Part 1. Hokkaido, Japan, Institute of Low Temperature Science, 1967, pp. 469–478.

Zotikov, I. A., V. S. Zagorodnov, and J. V. Raikovsky. Core drilling through Ross Ice Shelf. *Antarctic Journal of the United States*, Vol. 14, No. 5, 1979, pp. 63–64.

Zotikov, I. A., V. S. Zagorodnov, and J. V. Raikovsky. Sea ice on bottom of ice shelf. *Antarctic Journal of the United States*, Vol. 14, No. 5, 1979, pp. 65–66.

Index

Abele, Gunars, 161
Adventure Network (ANI), 148–150, 182, 187
Adventurers, 190
Aircraft
   Bluebird (P2V), 75,78
   DC-6B, 187
   DC-10, 40, 132, 145, 147
   El Paisano (R7V), 46–47
   Globemaster, C-124, 73
   Helicopters, xi–xiii, 21, 27–29, 34, 36–39, 43, 47, 49, 51–54, 65, 67, 69, 73–75, 81, 98–99, 106, 115, 119–136, 141, 145, 155, 179
   Hercules LC-130, C-130BL, L-382G, 39, 46, 70, 96, 98, 105, 114, 124, 129, 134, 137, 140–144, 150–152, 154, 158–159, 171–172, 175–176, 180, 184, 187
   Little Horrible (R4D) 53, 57, 61
   Marlene (R4D), 98
   Neptune, P2V, LP-2J, 27, 75, 78, 164
   Otter, DHC-3, UC-1, 12–13, 17, 27, 30–31, 49, 81, 102
   Pegasus (C-121), 165
   Phoenix (C-121J), 104, 108
   Porter, 102
   Que será será (R4D) 28
   Skymaster, DC-4, C-54, 27, 149–151, 159, 182, 187
   Skytrain R4D, LC-47, DC-3, 16–17, 21, 27–28, 35, 57, 69, 76–77, 83, 91–92, 96–99, 103, 106
   Starlifter, C-141, 121
   Super Constellation C-121J, R7V, 27, 31, 46–47, 103–116, 165
   Tri–Turbo DC-3, 148
   Twin Otter, DHC-6, 147–150, 153, 157–160, 163–167, 171, 175, 179, 181–184
   Wilshie Duit (R4D), 76
Air New Zealand, 132
Airdrop Peak, 61
Alaska, 73, 165, 172
Aldrich, Mount, 52
American Alpine Club, 150
American Antarctic Mountaineering Expedition, 150, 161

Amundsen Glacier, 83, 89–92, 100, 110
Amundsen, Capt. Roald Engebreth Gravning, 7, 10–11, 25, 43, 83, 89–90, 93–96, 99, 101
Amundsen-Scott Station (*see* South Pole).
Andersen, Dr. Bjørn G., 121
Ann Arbor, Michigan, ix, 1, 3, 46, 71, 93
Antarctic
   Airways, 161
   Peninsula, 102, 148, 185, 191
   Treaty, 4–5, 53, 95, 156, 189–191
Arctic, 5, 73, 102, 126, 148
Argentina, 4, 189, 191
Atkinson, Dr. Richard C., 140, 143
Austen, Jane, 27
Australia, 4, 7, 77, 106, 184, 186, 191
Austria, 191
Axel Heiberg Glacier, 83, 93, 96

Back, June D., 55
Bacon, Francis, 147
Badger, Thomas, 68
Bailey, Jeremy T., 116
Baldwin, Deborah G., 156
Balloons, xi, 64
Bamber, Dr. Jonathan, ix
Barne Inlet, 34, 51
Barne, Lt. Michael, 34
Barometers, 74–75
Bates, James, 30
Bayad, Col. J. K., 181
Beacon Heights, 154
Beardmore Glacier, 35–36, 57–58, 60–65, 74–75, 79–81, 85, 89, 110–111, 164, 166, 175–178, 180–181
Beardmore (weather station), 35, 53, 57–58, 61, 63–65, 67, 77, 79, 83, 91
Behrendt, Dr. John C., 142–143
Belgium, 4, 106, 191
Beltramino, Juan Carlos M., 191
Benson, William, 138
Bentley, Dr. Charles R., 117
Berg Field Center, 122,
Berkner Island, 143
Berry Peaks, 166

Betty, Mount, 93–95
Binder, Lt. Raymond A., 104
Black Island, 29, 31, 34, 36, 154
Blackburn, Quin A., 85, 99
Blaisdell, Dr. George L., 191
Bogard, D. D., 161
Bogorodsky, V.V., 117
Bolt, Lt. Ronald L., 81
Bolton rations, 15, 23, 38, 45
Bonney, Lake, 112
Bowers, Lt. Henry Robertson, 28, 54,
Bowie, Dr. Glenn E., 99
Brazil, 189, 191
Brecher, Dr. Harold Henry, 120–121, 125, 131, 144–145, 187
Bresnahan, David M., 135, 154
Britannia Range, 51, 126
British
    Antarctic Survey (BAS), xvii, 101–102, 104, 106, 119, 147
    Commonwealth Trans–Antarctic Expedition 1955–1958 (*Theron, Magga Dan, Endeavour*), 4, 31, 73
    Expedition 1901–1904 (*Discovery*), 3, 19, 23–24, 33–36, 43, 65, 113, 135, 158
    Expedition 1907–1909 (*Nimrod*), 11, 19, 36, 43–44, 51, 81, 181
    Expedition 1910–1913 (*Terra Nova*), 11, 28, 34, 43–44, 81, 93, 95, 99, 154, 156, 181
    Foreign Office, 53
    Imperial Trans–Antarctic Expedition 1914–1917 (*Endurance*), 11, 19, 62–63, 135
    Trans–Globe Expedition 1980–1981) (*Benjamin Bowring*), 147, 161
Bronston, Alan, 168
Brotherhood, Dr. John, 105
Brown, Arthur J., 157
Brown Peninsula, 154
Brudie, Raymond E., 169
Brunk, Dr. Karsten, 161
Brunt Ice Shelf, 117
Bryanston School, xiv
Buckley Island, 81
Buchwald, V. F., 184
Bulgaria, 191
Bull, Dr. Colin, ix
Bull Pass, 31, 112
Burke, David, ix, 82
Burma (Myanmar), xiv
Byrd
    Adm. Richard Evelyn, xiii, 4–7, 10–11, 24–25, 43, 46, 51, 69–70, 75, 83, 85, 87, 90, 99, 122–123, 136, 145, 175, 188

    Antarctic Expedition 1928–1930 (*City of New York, Eleanor Bolling*), 4, 6, 83, 85, 93–95, 99, 104, 134
    Antarctic Expedition 1933–1935 (*Bear of Oakland, Jacob Ruppert*), 67–85, 87, 99, 102
    Station, 7, 57, 115
    Glacier, xiii, 32–33, 37, 51–52, 54, 74–75, 99, 111, 119–134, 144
Caches (depots), 35, 62–63, 85, 92–95, 175–177, 181, 183,
Calgary, Alberta, 157
Callander, B. A., 191
Cambridge, England, vii, ix, xvi, 1, 5, 101, 104, 109, 119, 178
Cameron, Andrew O., 138
Cameron, Dr. Richard L., ix, 138, 187
Campbell, Richard J., 157
Canada, xv, xvii, 44, 46, 102, 191
Carbon monoxide, 67–69
Carlson, Lt. Ronald F., 35, 92
Cassidy, Dr. William A., 156, 165, 171–172, 174–175, 177, 181
Central Intelligence Agency, 191
Chalet, USARP, 121–122, 154
Chapman, William H., 81
Cherry-Garrard, Apsley, 28, 40, 156
Chichester, Douglas C., 168, 170–171
Chile, 4, 106, 149–150, 183, 189, 191
China, 186, 189, 191
Chipman, Elizabeth, 123, 130
Chocolate, Cape, 34
Christchurch, New Zealand, 3–5, 30, 40, 46, 70, 73, 104–105, 121, 143, 153–154
Christmas, 16, 64, 90, 94, 115
Clarke, R. S., 184
Claytor, Graham, 140, 143
Cleveland, Wisconsin, 5
Cloudmaker, The, 81, 177–178, 181
Clough, Dr. John W., 140
Coats Land, 117
Cockerell, Sir Christopher, 165
Cocks, Mount, 54
Colbeck, Lt. William R., 158
Coleman stove, 14, 38, 51
Colombia, 191
Colorado Glacier, 166
Compass, 77–78, 159
Contreras, Alejo Contreras Staeding, 181
Cooper, Paul R., 191
Coral, 48
Corner Camp, 19
Corr, Jerry, 181
Cosmology, 163

Coughran, William A., 158–159, 168
Crampons, 79, 85
Crary, Dr. Albert Paddock, 3, 11, 35, 49, 55, 93, 99, 103, 116, 123–124, 187, 191
Crockett, F. E., 93–94
Crozier, Cape, 27, 30, 37–38, 52–54, 113
Cryoconite, 95
Cuba, 191
Czarniecki, Louise A., 165
Czechoslovakia, 191

Dailey Islands, 34, 47
Dale, Lt-Cdr. Robert Lee, ix, 27, 35
Darby, Dr. David G., xi–xiii, 42, 46–50, 52, 55, 64–69
Darlington, Jenny, 123, 130
Darlington, Henry, 123, 130
Dartmouth College, 42
Darwin Glacier, 37, 124–134
David Glacier, 107
Dawson, Lt. Col. Merle R., 13, 16
Debenham, Professor Frank, 34–36, 48–49, 55
Debenham Glacier, 34
Deception Island, 106
Dehydration, 60
DenHartog, Dr. Stephen L., 168, 172, 175, 177–181, 191
Denmark, 191
Denton, Dr. George H., 125, 144
Depots (*see* Caches)
Dewart, Dr. Gilbert, 101, 116
Discovery, Mount, 31, 36
Doake, Christopher Samuel McClure, 191
Dogs, vii, xv, 21, 24–25, 42–43, 45, 60, 73–74, 85, 91–92, 94, 154, 180, 187
Dominion Range, 81
Dorrer, Dr. Egon, 191
Dreschhoff, Dr. Gisela A. M., 165
Drewry, Dr. David John, 141, 144, 191
Dromedary, Mount, 154
Dronning Maud Land (*see* Queen Maud Land)
Dufek, Admiral George J., 119, 123
Dufek Massif, 143
Durham,
    Mount, 85, 90
    New Hampshire, 85
    Point, 85, 89, 99

Ecuador, 191
Ellis, Murray R., 30
Ellsworth station, 7
Ellsworth Mountains, 148, 150
Elsner, Dr. Robert W., 165

Erebus
    Crash of Air New Zealand DC-10, 40, 132, 145, 147
    Mount, 19, 40, 106, 132, 145, 147, 154, 183
    Glacier Tongue, 19, 112
Eugster, O., 161
Evans,
    Cape, 28, 154
    Ice Stream, 187
    Dr. Stanley, 102, 104, 117, 187
    PO Edgar, 156
Evteev, Dr. Sveneld [Yevteyev], 30, 36–40, 52

Faiks, Senator Janice O., 172
Fastook, Dr. James L., 120, 133, 145
Ferrar Glacier, 35
Ferrar, Hartley T., 35
Fiennes, Sir Ranulph Twistleton-Wykeham-Fiennes, 147, 161
Filchner Ice Shelf, 139, 143
Filho, L. G. Meiro, 191
Finch, Lt. Jerry L., 104
Finland, 191
Fish, 47–48
Flags, 10
Ford, Dr. Arthur B., 143
Foster, Thomas H. II, 165
Fowler, Capt. Alfred N., 121
Fox, Adrian J., 191
Framheim, 7
France, xv, 4, 186, 191
Francis, Harry, 45
Frigon, Anthony, 159, 166–168
Frostbite, 15, 35
Fuchs, Sir Vivian Ernest, 4, 25, 31, 40, 73, 82
Fuerza Aérea de Chile (FACH), 149–150

Gamburtsev Mountains, 114
Gap, The, 61–62
Gateway, The, 62
Gausseren, Lt. William, 125
Georgia Institute of Technology, 134
Germany, 106, 186, 191
Giaever, John Schjelderup, xviii
Giovinetto, Dr. Mario B., 116, 191
Gneiss Point, 34
Goldie, Cape, 58
Goodale, Edward E., 3, 46, 93–94, 104
Gould, Dr. Laurence McKinley, 11, 25, 75, 83, 85, 89, 93–96, 99
Gow, Dr. Anthony J., 5–6, 55
Gowdy, Elsie J., 168, 172
Granite Pillars, 81

Gravity, 72–99, 102, 185
Great Lakes Naval Hospital, 45
Greece, 191
Greene, Lt. John H., 81
Greenland, 5, 73, 123, 148
Greenpeace, 23, 154–155, 188
Gressitt, Dr. J. Linsley, 5–6, 28, 31, 36
Guatemala, 191
Gudmandsen, Dr. Preben E., 117
Gushwa, John B., 169–171
Guthridge, Guy, ix

Hackerman, Dr. Norman, 140, 143
Hallett Station, 7, 106
Halley Bay, 105–106
Hamilton,
    Massachusetts, 93
    Mount, 83, 86–89
    Stuart, 181
Harold Byrd Mountains, 166
Harris, N., 191
Harvey, Ralph P., 177
Hattersley-Smith, Dr. Geoffrey Francis, 102
Havola, Major Antero, 5–6, 15
Headland, Robert Keith, ix, 191
Herbert, Wally, 73, 92, 96, 100, 123, 130
Hercules Inlet, 182
Heritage Range, 149
Hickey, Lt. John, xi, xiii, 49, 52–54, 69, 74, 81–82, 187
Highjump, Operation, 7
Hillary, Sir Edmund Percival, 4, 10, 25, 30, 40, 82, 132
Hofmann, Dr. Walther, 191
Holdsworth, Dr. Gerald, 112
Holmes and Narver Inc., 124
Hope, Mount, 36, 61–64, 69, 74, 80–81
Horney Bluff, 51, 111, 126, 144
Houghton, John T., 191
Hovercraft, 165, 183
Howe, Mount, 160, 170
Huffman, Jerry W., 107
Hughes, Dr. Terence J., ix, 119–135, 143–145, 147, 166, 187
Hungary, 191
Hunt, CPO Glenn C., 104
Hunt, Sir John, 10
Hurtig, Robert, 159
Hut Point, 19, 98, 135
Huxley, Leonard, 93, 99
Hyland, Dr. Mark R., 120, 128, 133
Hypothermia, 90, 128

Icebergs, 4, 7, 9
Ice drilling, 115–116, 140–141
Ice runways, 149–152, 156–157, 160, 166, 169, 175–177, 184
India, 191
Inertial navigation, 140, 159
International Geophysical Year (IGY), 4
Ipswich, Massachusetts, 93
Italy, 191
ITT Antarctic Services Inc., 153

Jacobi, Frederick J., 5–6, 11–15
Jamesway huts, 21–22, 35, 58, 67, 95, 124, 135, 141, 157
Jankowski, Dr. Edward A., 191
Japan, 4, 191
Jarina, Lt–Cdr. Michael, 98
JATO rockets, 35, 57, 75–76, 92
Johnson, P., 161
Jones, Sarah, 165
Jones, Dr. Thomas O., 3
Jordan, Susan R., 191
Joyce, Ernest Edward Mills, 25
J-9 ice drilling camp, 140–141

Kainan Bay, 3, 7, 9, 185
Kattenburg, A., 191
Kellogg, Dr. Thomas B., 135–136, 145
Kellogg, Dr. Davida E., 135–136, 145
Kenn Borek Air, 157
Kennedy, Nadene G., ix
Kerr, Richard A., 191
Kershaw, John Edward Giles, 147–150, 187, 191
Keys, Henry, 153,
King George Island, 189
King, William, 124–131
Kinnaird, Robinson S., 158
Knibbs, Capt. L. Brydon, 157, 159–160, 164–168, 171
Knoll, The, 28
Koerwitz Glacier, 166
Koettlitz Glacier, 34
Kon-Tiki Nunatak, 66
Kooyman, Dr. Gerald L., 165
Korea (PDR), 191
Korea, Republic of, 189, 191
Kovacs, Dr. Austin, 161
Krebs, Cdr. Manson, xi–xiii, 12–13, 36, 49–51, 70, 74, 187
Kyffin, Mount, 60–61, 79, 81, 177, 181
Kyle, Dr. Philip R., 135

La Count, Ronald R., 154–155, 163–164
Laird, FO Denis, 130, 133
Lakenheath airfield, 104
Larsen Ice Shelf, 191
Laws, Dr. Richard Maitland, 191
Leinmiller, Mark W., 134
Lennox-King, Lt–Cdr. James, 28–30
LePage, Carolyn A., 135
Lewis Cliff Ice Tongue, 165
Liestøl, Dr. Olav, 57–65
Likely, Wadsworth, 49, 51
Linder, Dr. Harold W., 72, 76–79, 83, 89, 91–93, 97–98, 187
Lingham, Jeffrey M., 120
Lister, Dr. Hal, 71
Lyttelton, Cape, 65, 77
Little America, 4, 6–8, 10, 15, 20–21, 75, 94, 102, 123, 177
Liv Glacier, 84, 95–97, 99–100, 110
London, xiv, 53, 78
Long, Jack B., 12–20, 23, 39, 187
Lucchitta, Dr. Bärbel K., ix

MacDonald, Capt. Edwin, 8, 10
Macdonald, John, 145
Macfarlane, Stuart, 145
Magnetometer, 77, 142–143
Mahon, Judge Peter. T., 145
Maine, 119–120, 125, 135
Mammoth Mountain Inn, 122, 155
Mañana Traverse Party, 39
Marie Byrd Land, 95, 116
Markham, Sir Clements Robert, 161
Marvin, Steven E., 157
Maskell, K., 191
Matthei, Gen. Fernando Matthei Aubel, 149–150
Maxwell, Dr. Michael, 149
McGregor, Dr. Vic R., 100
McGuire, PO Michael, 104
McKinley, Capt. Ashley C., 83, 93
McKinley, Capt. Brian, 159–160, 163–168
McMurdo Ice Shelf (pinnacled ice), 29, 36, 47, 135, 144–145
McMurdo station, 3, 6–7, 18–25, 27, 30–31, 34–35, 38–40, 42–43, 45–46, 49, 51–52, 54, 63–65, 69–71, 73, 75, 78, 91, 96–99, 102–109, 111, 114–116, 119–124, 130, 135–137, 143, 153–157, 163–166, 171–172, 180–181
McMurdo Sound, 3, 19, 31, 34, 36, 63, 105, 112–113, 135–136
Mellor, Dr. Malcolm, 151, 153, 165, 168, 180–181, 183–184, 187, 191
Mervis, Jeffrey, 191

Meteorites, 156–157, 161, 165, 170–172, 174–175, 177, 181, 184
Metz, Shirley, 181
Meyer Desert, 178, 181
Meyer, Dr. George H., 178
Midlam, Steven, 168
Mill Glacier, 81, 165, 176–181, 184, 187
Milmarik, Ronald, 181
Milton, John, 1
Minerals, 188–189
Minna Bluff, 31, 39, 54, 57, 154, 161
Minnesota, 72, 74
Mirage, 67
Mirny, 77
Missildine, Lt. Ernest, 104
Montana, 171
Morgan, Cdr, 140
Morning, Mount, 54, 154
Morrison, Lt–Cdr. James K., 104, 107–115
Mountaineers, 120, 124, 150, 190
Moxley, Lt. Donald F., 31, 34
Mulgrew, Peter D., 30, 40
Mulligan, Dr. John J., 34–35
Mulock, Lt. George, 33
Mulock Glacier, 33, 37, 49, 98, 111
Mulock Inlet, 33
Munson, Capt. William H., 35
Murden, Victoria, 181
Murphy, Joseph, 181

Nansen, Dr. Fridtjof, 44, 55
Netherlands, 191
Neuburg, Dr. Hugo A. C., 5–6, 53
New Hampshire, ix, 42, 85, 151, 184, 187
New Zealand, 3, 4, 21, 28, 39, 45–47, 73–74, 96, 104–106, 121, 124, 130–132, 138, 153–154, 191
 Antarctic Research Programme (NZARP/NZAP), 7, 20–21, 24, 28, 73–74, 92, 96, 106, 135, 154
Nichols, Mount, 166
Nimrod Glacier, 34, 36, 65–67, 71, 76–77, 111, 116
Ninnis Glacier, 115
Northwest Passage, xvii, 119
Norway, xv, xvi, 4, 44, 106, 121, 191
Norwegian Antarctic Expedition 1910–12 (*Fram*), 7, 11, 43, 83, 89–90, 93–96, 99
Norwegian-British-Swedish Antarctic Expedition, xv, xvi, 43, 55, 188
Nuclear power station, 73, 106, 122
Nyquist, L. E., 161

Oates, Capt. Lawrence Edward Grace, 156
O'Brien, John S., 93–94
O'Brien Peak, 90

Observation Hill, 19, 73, 156
Okuma Bay, 75
Olson, E., 184
Olson, Dr. James J., 72–98, 187
Operation Deep Freeze, 6, 24, 53, 95, 123
Otto, Richard V., 164
Otway, Peter, 82
Oxford, England, xiv–xvi

Paine, Stuart D., 85,
Palmer Land, 142, 148
Palmer Station, 106, 141
Papua New Guinea, 191
Parent, Mark J., 177–181
Patagonia, 124
Patriot Hills, 149–151, 154, 182–183, 187
Peary, R-Adm. Robert Edwin, 123
Pemmican, 58–59, 90
Penguins, 10, 12, 27, 30, 38, 42, 53–54, 113, 165
Pensacola Mountains, 151, 161
Peoples, Mary Ann, 156
Perk, Henry, 182–183
Peru, 191
Pesce, Cdr. Victor L., 140
Petrie, David L., 102
Pfeffer, Dr. William, 120, 133, 145
Pinnacled ice, (see McMurdo Ice Shelf)
Pinochet, General Augusto, 150
Plateau station, 106
Plunket Point, 176, 180
Point Mugu, California, 121
Polair, 148,
Poland, xiv, 186, 189, 191
Pole of Inaccessibility, 113
Port Hueneme, California, 5
Protocol on Environmental Protection, 190
Psychiatry, 45
Psychology, 45
Punta Arenas, Chile, 150, 157, 182–183

Queen Maud Land, xvi
Queen Maud Mountains, 75, 83, 93–94, 96, 99, 110

Radcliffe, Peter, 124–129, 133
Radio echo sounding, 102–117, 136, 141–147, 185
Raikovsky, Dr. J. V., 145
Raney, Dr. Michele E., ix, 138
Rations, 15, 23, 30, 38, 45, 58–60, 74, 122, 132
Reagan, President Ronald, 191
Reece, Dr. Alan, xvi, 45, 55
Reedy, Adm. James R., 123, 163
Reedy Glacier, 110, 163, 166
Renirie, Jack, 158–159

Reston, Virginia, ix, 152
Richards, R. W., 63, 70
Ridley, Julian B., 158
Riley, Lt. Stephen G., 104
Roberts, Dr. Brian Birley, 53–54
Robin, Dr. Gordon de Quetteville, ix, 101–117, 136, 141, 187, 191
Robinson, Dr. Edwin S., 116, 191
Romania, 191
Ronne, Cdr. Finn, 123
Ronne Ice Shelf, 181
Roots, Dr. Ernest Frederick, 43
Rose, Lisle, 136
Ross
    Capt. Sir James Clark, 9, 11, 19, 25
    Ice Shelf, 1, 3, 7–8, 11–21, 24, 35–37, 49–50, 58, 69, 75, 93, 99–100, 102, 110, 128, 134–136, 140–41, 144–145, 164, 178, 181, 185, 191
    Island, 3, 19, 27–28, 52, 112, 145
    Sea, 2–3, 11, 94, 135, 153
Royal
    Air Force, xv, 17
    Navy, xv, 5, 14, 24, 33
    New Zealand Air Force, 130, 154
    New Zealand Navy, 28
    Society, 101
    Society Range, 34, 54
Royds, Cape, 135
Rüfli, Henry, 121
Rundle, Dr. Arthur S., 71–99, 110, 160, 187
Runways on ice, 147–184
Russell, Dr. R. Scott, xv
Russell, Richard, 85, 87
Russia, 186, 191
Russians, 23, 30, 36, 101, 114, 140–141, 156

Salisbury, Mount, 166
Sandefjord, Norway, xvi
Santiago, Chile, 149
Sastrugi, 76, 85, 90, 113, 175
Schroeder, James E., 5–20, 187
Schutt, Dr. John W., 177
Schytt, Dr. Stig Valter, xv, 161
Scientific Committee on Antarctic Research (SCAR), 189
Scott
    Base, 19–21, 28, 30, 73, 96, 130–132, 154
    Glacier, 83–89, 100, 110, 160, 166
    Polar Research Institute, ix, xvii, 99, 101–102, 116, 136, 141–142
    Capt. Robert Falcon, xi, xii, xvii, xviii, 3, 11, 19, 23–25, 33–34, 40–41, 44, 51, 54, 60, 70, 81, 85, 90, 93, 95, 112–113, 131, 161, 174

Sealdrum, 72, 176, 181
Seals, 38, 48, 135
Seaver, George, 40
Seelig, Walter R., 121
Seismic sounding, 39, 72, 75, 77, 79, 89–90, 92, 96, 99, 102, 111
Serson, Dr. Harold, 44
Seufert, Wilfried, 191
Shackleton
    Edward, 55
    Ernest Henry, 11, 19, 25, 36, 40, 43–44, 51, 54, 60–65, 70, 81, 83, 90, 163, 176–177, 181
    Glacier, 67, 176
    Inlet, 65
Shakespeare, William, 185
Sharp, Michael, 181
Ships
    Arneb, USS, 5, 6, 8, 10, 40
    Atka, USS, 7, 9, 39–40, 114
    Discovery, RYS, 135
    Dreadnought, HMS, 119
    Endurance, RYS, 63
    Erebus, HMS, 9
    Glacier, USS, 8
    Kapitan Belousov, 40
    Labrador, HMCS, xvii
    Manhattan, SS, 119
    Norsel, MV, xvi
    Southwind, USCGC, 40
    Terror, HMS, 9
    Thorshøvdi, MV, xvi
Shipton, Eric, 124
Shirtcliffe, James, 105
Shuman, Capt. Edwin, 8
Signor, Gary, 68
Silverstein, Samuel C., 161
Siple, Dr. Paul A., 24–25, 46, 54, 130, 134, 145, 180, 184
Siple Coast, 75. 158, 164, 188
Skelton Glacier, 31, 49, 111
Skvarca, Pedro, 191
Sladen, Dr. William Joseph Lambart, 113
Sloan, Valerie Jo, ix, 177–181
Smith
    Dr. Beverley Michael Ewen, 103–117
    Capt. James, 150
    Capt. John, 157
    Philip M., 75, 92, 104
    Thomas, 120, 131
Snakeskin Glacier, 175–176, 181
Snowmobiles
    Eliason, 44, 46, 57–69, 72, 89
    Polaris, 72, 79–97

Sno-cat, 3, 10–14, 16, 20–21, 24, 39
    Skidoo, 156–157, 167
Snyder, Lt. Steven, 52, 69
Socks Glacier, 81
Socha, Dr. David Grimes, ix
Søndergaard, Finn, 142
South Africa, 4, 106, 143, 148, 191
South Pole (and Amundsen-Scott Station), xvii, 2, 7, 28, 36, 39, 42, 59, 62, 67, 83–86, 89, 92–96, 99, 105–106, 114, 116, 124, 128, 130–131, 134, 137–140, 142–143, 151–153, 157–160, 163, 165–166, 168–174, 176–178, 180–184, 187, 190–191
South Shetland Islands, 189
Soviet Antarctic Expedition, 101, 113, 178
Sovetskaya station, 113
Spain, 191
Spitsbergen, xv, 73
Splettstoesser, Dr. John F., 125
Sponges, 47–48
Staiger, Rudolf, 161
Stark, Peter, 191
Stephenson, Dr. Simon N., 164
Stetz, Lt. Elias, 75, 78
Stonington Island, 123
Stuiver, Dr. Minze, 145
Sullivan, Lt. John, 129, 133
Sullivan, Walter, 130
Supporting Party Mountain, 75
Sweden, xv, 191
Swithinbank, Mary, ix
Swithinbank Range, 187
Switzerland, 121, 191

Tate, Kay Fogg, ix, 169–172
Taylor
    CPO Robert, 104
    Edith, ix
    Glacier, 112
    Thomas Edwin, ix, xi, 43, 46–110, 160, 181, 187
    Valley, 112
Telluride, Colorado, 43
Tellurometers, 81, 129, 185
Temperatures
    Air, 16, 28, 31, 35, 40, 47, 50, 67, 76–77, 80, 89, 93, 109, 113, 116, 126, 128, 131, 135, 139, 150, 154, 158, 160, 167, 178
    Ice, 50, 67, 89, 93, 116
Tennant, Lt. John W., 129–130
Tents, 38, 44–45, 50–51
Territorial claims, 5, 95, 189
Terror, Mount, 19
Theodolites, 14, 18, 38, 43, 47, 50–52, 63–64, 73, 87–89, 98, 127, 129

Thiel, Dr. Edward C., 72–78, 80, 99, 102, 187
Thiel Mountains, 183, 187
Thorne Glacier, 85
Thorne, George A., 93–94
Toney, George R., 20, 27, 65
Tourism, 189–191
Transantarctic Mountains, 24, 31, 33, 35, 43, 54, 75, 110–111, 119, 140, 154, 156
Tuck, Lt. (j.g.) John Jr., 42, 51–52, 57, 65, 67–68, 187, 191
Tyree, Adm. David Merrill, 24–25, 43, 73, 102, 163, 191

Ukraine, 191
Universities
    Arizona, 10
    Cambridge, 101
    Dartmouth College, 42
    Durham, 71
    Grand Valley State College, 10
    Maine, 119, 135
    Michigan, ix, 1, 42, 71, 93, 101, 126, 144
    Nebraska–Lincoln, 10
    Pittsburgh, 156
    Southern California, 10
    Southern Methodist, 10, 135
    Technical University of Denmark, 142
    Wisconsin, 140
United States
    Air Force, 39
    Antarctic Program (USAP/USARP), 21, 24, 45, 65, 101, 103–104, 106, 121–122, 132, 147, 151–158, 181–182, 185–187
    Army Cold Regions Research and Engineering Laboratory (CRREL), ix, 151, 153, 156, 161, 168, 184, 191
    Army Signal Corps, 102
    Board on Geographic Names, 85, 91, 187
    Central Intelligence Agency, 191
    Coast Guard, 40, 155
    Department of State, 95, 101, 136
    Geological Survey, ix, 43, 120–121, 142, 152, 168, 191
    National Aeronautics and Space Administration (NASA), ix, 164
    National Science Board, 135, 140, 143
    National Science Foundation (NSF), ix, 3, 11, 20, 27, 41–42, 45, 73, 75, 81, 104, 107, 116, 121, 123, 135–136, 138, 141–142, 151–154, 158, 165, 172, 191
    Navy, ix, xi, 5–10, 12–13, 16–17, 20–21, 28, 35–36, 39, 43, 46, 57, 64–65, 68–69, 73, 81, 98–99, 104, 106, 110, 115, 124, 140, 143, 154–155, 157, 184, 186, 188
Uruguay, 189, 191

Vanda, Lake, 31
Vaughan, David G., 191
Vaughan, Norman D., 93–94
Vette, Gordon, 145
Victor, Paul Émile, 123, 130
Victoria Valley, 31, 112
Vida, Lake, 31
Vinson Massif, 148–150, 190
Vostok station, 113–114

Wade, Mount, 58, 83
Waite, Amory H. Jr., 102
Washington DC, 3–5, 41, 73, 101, 103, 141, 150, 186
Weasel tractor, xvi
Weddell Sea, 63, 139
Weeks, Lt. James W., 17, 76, 83
Weinstein, Daniel C., ix
Wells, Robert L., ix
Whales, Bay of, 3, 7, 85
Whaling, xvi, 38
Whiteout, 16, 61, 69, 90–91, 105, 124, 127, 132, 145
White Island, 31, 154
Whitmill, Leland D., 120
Wilkes Station, 7, 77–78, 138
Wilkniss, Dr. Peter E., 172
Williams, James, 181
Williams, Martyn, 181
Wilson
    Cape, 36, 77
    Charles R., 55
    Dr. Edward Adrian, 27–30, 36, 54, 65,
    Dr. Ove, 165
    Piedmont Glacier, 31, 34
Wirdnam, Sq-Ldr. K. A. C., 17, 34
Wisconsin Range, 163
Wohlschlag, Dr. Donald E., 53, 55
Women, xv, 119, 122–124, 138–139, 154, 181–182
Wordie Ice Shelf, 191
World War II, xv, 40, 43, 77
Wren, Christopher [Jr.], 71
Wright Valley, 31

Zagorodnov, V. S., 145
Zeller, Dr. Edward J., 165
Zotikov, Dr. Igor A., 49, 55, 140–141, 145
Zumberge, Dr. James Herbert, ix, 1, 3–13, 36, 135, 187, 191